BEFORE THE REFRIGERATOR

How Things Worked
Robin Einhorn and Richard R. John, Series Editors

ALSO IN THE SERIES:

Sean Patrick Adams, *Home Fires: How Americans Kept Warm in the Nineteenth Century*

Ronald H. Bayor, *Encountering Ellis Island: How European Immigrants Entered America*

Bob Luke and John David Smith, *Soldiering for Freedom: How the Union Army Recruited, Trained, and Deployed the U.S. Colored Troops*

David R. Danbom, *Sod Busting: How Families Made Farms on the Nineteenth-Century Plains*

Phillip G. Payne, *Crash! How the Economic Boom and Bust of the 1920s Worked*

Sharon Ann Murphy, *Other People's Money: How Banking Worked in the Early American Republic*

Johann N. Neem, *Democracy's Schools: The Rise of Public Education in America*

Benjamin F. Alexander, *The New Deal's Forest Army: How the Civilian Conservation Corps Worked*

Before the Refrigerator

HOW WE USED TO GET ICE

JONATHAN REES
COLORADO STATE UNIVERSITY–PUEBLO

Johns Hopkins University Press | *Baltimore*

Printed in the United States of America on acid-free paper
9 8 7 6 5 4 3 2 1

Johns Hopkins University Press
2715 North Charles Street
Baltimore, Maryland 21218-4363
www.press.jhu.edu

Library of Congress Cataloging-in-Publication Data

Names: Rees, Jonathan, 1966– author.
Title: Before the refrigerator : how we used to get ice / Jonathan Rees.
Description: Baltimore : Johns Hopkins University Press, 2018. | Series: How things worked | Includes bibliographical references and index.
Identifiers: LCCN 2017037463| ISBN 9781421424583 (hardcover : alk. paper) | ISBN 1421424584 (hardcover : alk. paper) | ISBN 9781421424590 (pbk : alk. paper) | ISBN 1421424592 (pbk : alk. paper) | ISBN 9781421424606 (electronic) | ISBN 1421424606 (electronic)
Subjects: LCSH: Ice industry—Technological innovations—United States. | Ice industry—United States—History.
Classification: LCC HD9481.U5 R44 2018 | DDC 338.4/7621580973—dc23
LC record available at https://lccn.loc.gov/2017037463

A catalog record for this book is available from the British Library.

Special discounts are available for bulk purchases of this book.
For more information, please contact Special Sales at 410-516-6936 or specialsales@press.jhu.edu.

Johns Hopkins University Press uses environmentally friendly book materials, including recycled text paper that is composed of at least 30 percent post-consumer waste, whenever possible.

CONTENTS

PREFACE: WHY ICE?

General Electric introduced the first modern electric household refrigerator in 1927.[1] Today, 99.5 percent of American homes own or have access to this particular appliance.[2] Much of the perishable food that stocks those refrigerators gets handled as part of the supply chains that bring those goods from their point of production to their point of purchase. Most people in the United States (and in many other developed countries around the world today) cannot live without a refrigerator or the benefits of refrigeration because they have such a positive impact on their lives. Refrigeration introduces the possibility of a healthier and more varied diet, saves time for whoever prepares food in the household, and lowers food costs by cutting waste. People in developing countries around the world generally crave refrigerators as soon as they get access to a reliable electrical grid because the benefits of refrigeration are important attributes of modern living.

The forerunner of the refrigerator and refrigerated supply chains of today was the American ice industry. The natural ice industry began in the early nineteenth century when Boston merchant Frederic Tudor started cutting ice from New England lakes, streams, rivers, and ponds. Tudor first created the demand for his product by convincing patrons in bars and taverns to drop ice into their alcoholic drinks. That habit gradually spread to other beverages as ice became cheaper and more accessible. Meatpackers began experimenting with ice refrigeration to ship their products as early as the 1860s.[3] Starting around 1880, large, bulky ice machines the size of small houses appeared on the American scene and gradually became an important source for the American ice supply.

By the early twentieth century, ice had become an essential tool for preserving perishable foods of all kinds. "Ice enters into dozens of phases of our daily life," explained a reporter for the *Salt Lake Herald-Republican* in 1909. "The butchers, grocers and others who sell edibles use ice to keep them. In iced, or refrigerator cars fruits and vegetables, meats, milk, oysters and many other things are shipped from one part of the country to the other. . . . All over the country there are cold storage warehouses where eggs, poultry and meats are sometimes kept for two years."[4] Consumers also bought and stored ice in boxes to protect perishable products like these in the early nineteenth century. Those iceboxes were gradually replaced by electric household refrigerators starting in the 1920s.

As the amount of and uses for ice increased, it became the centerpiece of an elaborate infrastructure that preserved perishable foods on long journeys from the country's most productive farmlands to consumers' kitchens around the country. Eventually, refrigerating engineers came to call these routes between the point of production and the point of consumption for perishable foods "cold chains." Cold chains are similar to what most people know as food chains, only they exclusively involve foods that require refrigeration during transport. Cold chains were (and remain today) a series of related technologies that worked together toward bringing perishable food (including ice itself) from the point of production to the point of consumption. Taken together, these many cold chains came to form a "seamless web" for delivering ice and other perishable products to the masses.[5]

Despite the important role of ice in daily life in America, few historians (besides me) have spent much time closely examining the ice industry.[6] Descriptions of ice instruments and harvesting practices were fairly common in newspapers during the ice industry's heyday because so few people understood how their ice got to them. The descriptions in this book of how ice was harvested, stored, and transported are intended to allow these practices to be understood and appreciated for their long-term historical significance. That significance derives from their influence on what came later. By establishing a market for perishable foods, ice laid the groundwork for the modern mechanical refrigeration–centered distribution system that followed. Even if the technological components of the cold chains of the pre-mechanical refrigeration era no longer exist, their influence on how and whether perishable food reaches us today explains why they deserve consideration.

The ice industry also had an extraordinary impact on the types of foods that Americans ate. Ice turned the manufacture of perishable food products like meat and lager beer into year-round activities, thereby increasing their supply and decreasing their price. It improved the geographical reach of perishable food products and even made it possible for consumers to defy seasons for the first time, since cold storage warehouses (at first generally refrigerated by ice) were able to preserve items such as apples or eggs, which had particular seasons, until they could be distributed at whatever time during the year that consumers wanted them. Ice also improved the supply of perishable foods of all kinds simply by cutting spoilage, no matter what the season, which likewise drove down prices and improved accessibility to members of the lower and middle classes.

Yet the historical significance of ice transcends merely its influence on what Americans once ate. How we used to get ice can explain much about the interaction between technology and capitalism, that between technology and consumption, and how technology could alter the physical landscape to meet the needs of producers and consumers. Ice in some ways also came to define what it means to be an American because of its many unique effects on what we eat and how we live. The dietary changes that refrigeration brought increased happiness and improved the overall health of the population. All these changes began with the ice industry, not with the electric household refrigerator.

The web of cold chains which the ice industry created before the refrigerator was as elaborate as the electrical or water systems serving great American cities at that time. This web was also very efficient and technologically sophisticated. "The story of the growth and development of the use of ice parallels in many ways that of such products of inventive genius as the telephone, the electric light and the automobile," claimed a group of New England ice dealers in 1931.[7] While there's obviously more than a touch of marketing in a claim like that, it is still worth considering how difficult it was to provide ice to Americans even at that late date. Tools for producing, storing, and transporting ice had been developed, perfected, and employed in such a way that ice became affordable to enough people to justify the costs of developing this complex system.

To understand how this system developed requires an appreciation for how all the technologies along a cold chain interacted. In an electrical system, for example, lights will not turn on if one component of the system breaks down. This was not the way that the ice industry operated. If producers harvested less ice or if they failed to protect it in every step on its journey to the point of consumption, consumers could still get ice. With less outside ice to protect the inside ice or warm gaps in the cold chain during transportation, less would be around for actual use. However, the system still provided ice. As the system became more efficient, more uses for ice developed and more people could afford to purchase it and the perishable food products that it preserved. The many effects of this increasingly efficient system changed the lives of ordinary Americans forever.

This book considers the details of the technologies that made ice production, storage, and transportation possible, how the technologies that made up this system improved over time, and the many effects of these improvements on the people who benefited from these changes. While increased access to ice may seem like a small change to anyone who can get all the ice they want, anytime they want, just by opening their freezer door, this seemed like something of a miracle to the ice industry's first customers. Understanding how that miracle happened demonstrates how the interaction between technology and culture over time—as well as how the many benefits that those changes brought to Americans—affected people up and down the economic ladder. Even though refrigeration has changed since the heyday of the ice industry, the same essential benefits that ice once provided continue today.

BEFORE THE REFRIGERATOR

INTRODUCTION. A Museum for the Ice Industry

MOST INDUSTRIES LEAVE TANGIBLE LEGACIES that are visible in the products that they produce. The American steel industry, for example, has left railroad tracks, skyscrapers, and automobile bodies. The American automobile industry has produced cars for more than one hundred years now, many of which are preserved in museums across the country or driven by car enthusiasts around the world. Even the coal industry has left piles of refuse next to the many holes in the ground dug to supply energy for America's industrial revolution and beyond. Any kind of food production is different in that its end result is either consumed or destroyed because of spoilage. Its primary legacy went into the people it first fed rather than anything it built. With those people gone, all that remains are pieces of the infrastructure necessary for the production and distribution of their particular products. Old beer cans, for example, are interesting antiques. Old beer, on the other hand, is of no use to anyone.

Imagine that you wanted to build a museum dedicated to the American ice industry, focusing on the late nineteenth and early twentieth centuries.[1] You would immediately realize that you had a big problem on your hands—none of the product of that industry has survived because it, of course, melted. You might then think of filling your museum with the kinds of products that benefited from ice as a source of refrigeration. Unfortunately, none of those products has survived either, since the food or drink that the ice kept cold evaporated, was eaten, or eventually spoiled. The only thing left to keep in this museum would be the tools and machines that companies developed in order to transport, store, and eventually manufacture their product.

In 1880, most ice companies still cut ice from lakes and streams in northern climes and sent that ice all over the country. By 1930, production methods changed even though the uses of ice remained similar. At this later point, many small companies manufactured ice in factories, or people made their own ice at home in electrical household refrigerators. Sometimes the consumers who bought this ice noticed the changes that occurred because the ice they bought became more or less expensive. Sometimes the consumers who bought this ice did not know that changes were occurring because the actual ice they used didn't appear to change at all.

To illustrate the breadth of the technological changes in the ice industry, consider three possible technologies that we might include in our hypothetical ice museum. One tool could be an ice auger—simply an iron rod, two or three feet long, with a small section at the end protruding at a right angle. The rod is marked off in inches because harvesters used ice augers like these to measure thickness. First, a hand drill like those developed for ice fishing would be used to cut a hole in the ice. The auger would follow. The angled portion would hook against the bottom of the ice in order to ensure the exactness of the measurement.

The thickness of their product mattered a great deal to the natural ice industry. If harvesters cut ice when it was too thin, it might crack, and laborers and horses might fall into the water below. Even if it failed to crack, thin ice would have inevitably melted into a puddle before the summer came since there would not have been enough outside ice to keep the inner parts cold. If they cut the ice when it was too thick, then the resulting blocks could be too heavy to move. Harvesters in different places cut ice to different thicknesses. The colder it was, the thicker the ice these entrepreneurs preferred. Six inches was enough in Maryland, just about the southernmost portion of the region where ice could be harvested. In Maine, harvesters might wait and cut blocks that were as much as twenty or thirty inches thick. No matter where ice got cut, people used ice augers to measure it before they started.[2]

In summer, once the ice arrived near its intended destination, that ice had to be distributed by ice companies to the businesses and households that had signed contracts for it. This was the job of ice deliverymen. Icemen had to be large and strong in order to lift all that ice. Since ice melts, the amounts specified in those contracts were always measured by weight. However, sometimes people asked for more ice right on the spot. Sometimes people complained that their iceman had shortchanged them. Sometimes ice companies worried that their deliverymen were selling ice on the side, pocketing the proceeds and chalking the difference up to melting. After the evidence melted, ice companies had difficulty checking up on the honesty of their employees.

Coupon books were the solution to this particular ice-related problem. Each coupon book had a series of small pieces of paper which represented a popular amount of ice to buy, like twenty-five pounds per household. Customers would buy the coupon books directly from the ice company. This limited the opportunities of icemen to directly handle cash. Customers received a small discount for buying in bulk. Drivers had to create a detailed record of the ice they delivered on their routes, which not only cut down on fraud but also made accounting for this particularly perishable commodity much easier.[3]

The first successful artificial ice machine would have had a hard time fitting in any museum because of its huge size. For example, a 220-ton refrigerating machine from the De La Vergne Refrigerating Machine Company of New York, New York, actually weighed about five tons, but it got that name because it could produce 220 tons of ice each day for sale throughout the area of operation for the company that bought one. At first, most of these machines operated in southern parts of the country since it was much harder to get natural ice to those areas. However, as mechanical ice machines gradually became smaller, cheaper, and more efficient, these machines began to replace natural ice harvesting throughout the United States.

Eventually, commercial ice machines produced a steadier, more reliable cold than natural ice ever could. While other countries primarily used this kind of equipment to prepare food (particularly meat) for export, only the United States used mechanical refrigeration to satisfy the already-existing demand for ice. Artificial ice was superior in just about every way to the natural ice cut from waterways across the country because it was cleaner, colder, and easier to produce. While natural ice persisted until around World War I, improvements in the reliability and cost-effectiveness of the machine-made product meant that the United States could become the only country on Earth which produced enough ice that it became affordable to the masses.

It is possible to see the continuity and changes in the American ice industry just from these three objects. Ice augers remained an important tool for ice harvesters as long as a market existed for natural ice. Eventually, however, ice made by machine made that kind of ice making, and hence that kind of tool, obsolete. Ice delivery depended on coupon books whether that ice came from a pond or a factory. Large ice plants were incredibly disruptive with respect to the rest of the ice industry, but their heyday lasted only three decades, as Americans quickly came to rely on their electric household refrigerators to supply the ice to which they had grown accustomed.

If this imaginary ice museum were to focus solely on the early ice-harvesting days, it might stress the innovative spirit that turned something that was literally everywhere in New England during winter into something that could make the merchants who harvested this product rich. It might also emphasize the considerable marketing savvy needed to get Americans and people around the world to buy a highly perishable product that they never knew they needed or wanted before it became available. However, the market for America's ice industry during its early days was almost exclusively overseas. Therefore, the early American ice industry offers little insight into the relationship that the American consumer had with what is now a quintessentially American product.

By focusing on the later years of the American ice industry, the years when Americans produced far more ice than they did earlier in the nineteenth century, this study can help explain why the majority of Americans took to ice so quickly and why that love has persisted down to the present day. Over time, the quality of ice cut from lakes improved. Likewise, manufactured ice had many benefits over the natural ice that had made the industry possible in the first place. Because of these changes and the many improvements in producing, transporting, and storing ice, the uses of ice and the refrigeration that ice provided expanded rapidly over the course of the nineteenth century. This growth sprung not from a single invention, but from a vast array of improvements. The effects of those improvements were extraordinarily important to both what Americans ate and how Americans lived.

Much of this book could serve as the text for this ice museum, where it might accompany the exhibits that explain the relationship between the innovations that made all these changes possible and the vast array of consumers who benefitted from them. Chapter one looks at the late natural ice industry—after the domestic market was established and just as it started to face serious competition. That competition came from mechanical refrigeration, the subject of chapter two. It follows ice machines from their first awkward commercial versions through the many improvements that led to the near-complete elimination of its once-mighty harvested competitor.

Chapter three covers the many different ways that both natural and artificial ice was transported from the point of its production to wherever it (or the perishable food that it protected) would be consumed. To provide point-to-point delivery for something as highly perishable as ice required many different technologies and methods of transportation, even for relatively short cold chains. Chapter four considers the way that Americans used ice, both commercially and at home. This is the chapter of the book that focuses the most on how ice changed the American diet. In general, it is safe to say that the more ice that Americans could use, the more uses they found for it. Chapter five explains how household refrigeration gradually displaced the elaborate ice infrastructure by shortening the cold chain down to practically nothing. While there is still an ice industry in the United States, the market it serves is completely different from the one it served in the past.

But if this hypothetical museum existed, its visitors would learn a great deal more than just the many ways in which we used to get ice. For example, they would see that not all ice was the same. Ice cut from a lake was just as cold as ice made in a factory, but it was sometimes better suited for particular purposes than the cleaner, clearer artificial product. They would also see that different parts of the ice industry often developed many different ways to perform the same task, sometimes at the same time. Iceboxes and ice blankets both slowed the speed at which ice melted, but they did so for entirely different reasons. Both these tools demonstrate how sometimes just the simplest innovations could perform a particular job better than more complicated technologies.

Visitors to this ice museum could also see the important role that markets played in defining how all the improvements in the ice industry interacted toward the goal of creating more ice for everyone. Every part of the cold chain depended on an end point, consumers or businesses willing to purchase the ice or perishable foods that depended on ice and refrigeration for their existence. If changes between the point of production and the point of consumption affected the ice adversely, consumers sometimes changed their habits in response. Entrepreneurs and inventors sometimes built a better mousetrap, only to find that consumers did not see that change as better because of its effects on the cost or the taste of the product in question.

In other words, the success of the ice industry in the United States was not just a testament to American ingenuity. The ice industry also depended on the willingness of both producers and consumers to foster and accept change—sometimes extremely rapid change. The most successful of those changes increased the amount of ice and ice-related products available while simultaneously lowering their prices. The more ice that became available, the more ice Americans consumed. To this day, people in many other countries privilege the freshness of food over its abundance because they value taste over price. In the United States, our general tendency to accept anything chilled or frozen as fresh enough to meet our culinary needs dates from the success of the ice industry to grow the market for its product.

What separated this industry from so many other cutting-edge industries of the late nineteenth and early twentieth centuries was its wide array of actors. Even the largest ice companies of the time could not dominate local markets in the same way that John D. Rockefeller dominated the entire American oil industry or Andrew Carnegie could dominate the American steel industry for a time because the supply and demand for ice were so unreliable. A cold winter (in the right regions) would increase supply. A hot summer would increase demand, and matching supply and demand for unique local markets around the country was just too difficult for any one company to achieve. As a result, there was never one ice market during the late nineteenth and early twentieth centuries. There were lots of little ice markets, with distinct sources of supplies and distinct pockets of demand.

The fact that such a decentralized industry nonetheless grew so much demonstrates the important role that capitalism has always played in organizing distribution networks that have made the commodities that we depend on in our everyday lives available to everyone. Innovations spread freely between local markets because different actors were responsible for different technologies and the success of an ice company in the next state over did not hurt your local business. The promotion that local ice delivery companies did to increase demand for ice helped the ice harvesters they bought from, as well as the taverns that bought their ice. In fact, every enterprise along the cold chain worked toward the common goal of producing and preserving enough ice to do its appointed job in the seamless web of cold chains that every American eventually came to depend on.

This history of the American ice industry begins with the rise of mechanical refrigeration, around 1880, and follows it until its effective demise in 1930, continuing on to its complete eradication by 1960. America's countless refrigerators and the vast but seldom-recognized refrigerated infrastructure that supplies them are the closest thing we have to a physical legacy of the American ice industry apart from the scattered technological artifacts that might populate our ice museum. While the ice and refrigeration industries relied on very different kinds of innovations, they both were (or are) devoted to achieving the same goal—the preservation of perishable food. The more food the technologies preserve, the lower the price and the more of it people can eat.

In many ways, the cold chains of the late nineteenth and early twentieth centuries were an even more complicated achievement than the cold chains of today because the earlier version utilized so many different technologies, from ice augers to ice machines.[4] Simply keeping ice from melting prematurely required coordination between a vast network of people and enterprises. Which technologies they used depended on where their ice came from and what the customers who bought that ice expected to do with it.

By focusing on the prolonged interaction of innovations all along the cold chain, it becomes possible to show how improvements in refrigeration happened, and how those improvements eventually led to the demise of the ice industry itself, as refrigeration inevitably passed from natural to mechanical. While there may not be any museums devoted entirely to the ice industry, people in northern New England or Upstate New York sometimes remember the ice harvests that took place in those regions as a celebration of their heritage. A main goal of this book is to convince you that the dynamics of this industry still have much to teach us about how technologies change over time. This particular set of technologies also had everything to do with what we ate and when we could eat it, just as the mechanical refrigeration that eventually replaced the old ice industry does today.

1 How to Harvest Ice

WHEN THE BOSTON MERCHANT Frederic Tudor sent a shipload of ice to the Caribbean island of Martinique for the first time in 1806, he lost between $3,000 and $4,000 on the venture.[1] Over time, Tudor improved his methods of cutting, storing, and transporting all the ice he sold. Even though many competitors copied his methods and the tools he used to make ice harvesting into a viable business, few Americans understood how complicated getting ice to them could be. Harvested in the winter, stored until summertime, sometimes transported from thousands of miles away—the mature ice business depended on a long list of distinct instruments and practices, many of them gradually developed over decades.

Despite its accumulated knowledge, the ice industry in America still depended entirely on the weather to create its product for most of the nineteenth century. When the winter in the northern parts of the country was cold for long stretches, without any rain to weaken the product, the resulting ice was thick, clear, and easy to harvest. When it was generally warm, or even when a warm streak came at an inopportune time, ice companies had to deal with shortages. A season like that was called an "open winter," and the first such season that the industry faced came in the winter of 1879–80.[2] Unfortunately, this open winter stretched well beyond the New York market. "From our advices, the prospects are poor in Michigan, Ohio and other northwestern states," reported the Wheeling, West Virginia, *Daily Intelligencer* earlier that same year. "The universal dearth will make this commodity very scarce and very high next season, unless something in the shape of a solid freeze intervenes in the meantime."[3] It didn't.

Warm weather that winter meant high prices the following summer. While this had been a periodic problem since the beginning of the industry, what made this particular open winter so important was that the size of the natural ice industry had grown considerably around the country in preceding years. Now ice harvesters had to figure out how to satisfy the demand that they themselves had cultivated even though their supply had decreased sharply. Otherwise, many consumers might realize that they could live without ice—just as they had before the natural ice industry had appeared.

Yet the ice industry continued growing in the face of such hurdles because the natural ice industry changed after 1880. Ice harvesters harvested more ice from more places. They then sold that ice in local markets rather than overseas, as they had earlier in the nineteenth century. Since ice had a shorter journey going between locations in the same region of the country, less of it melted on the way to the people who consumed it. Another reason that the ice industry grew were the many new inventions that helped make it more efficient than it had been previously. Better tools were developed to help harvest the product. Parts of the harvesting process, like lifting the ice into the icehouse, became automated. Because of these changes, the natural ice industry began to take on many of the attributes of efficient modern cold chains. As ice machines and other innovations were developed, they could easily be adopted into the systems that the natural ice industry had pioneered.

New modes of transportation, like faster railroads and specially built barges, also helped get more ice to market faster. Again, traveling over shorter distances and at quicker speeds meant that less ice melted on its way to market. More ice generally meant lower prices, assuming that the market in ice was competitive (and that was not always the case). Lower prices drove up demand and eventually expanded the uses for ice beyond just dropping it into drinks. All these changes would have meant a large increase in the total amount of ice in America even if the mechanical ice machine had never been invented.

The increase in volume for the natural ice industry was staggering. In 1856, 146,000 tons of ice left Boston (the original transit point for the product of the New England ice industry), mostly for points overseas, but also for a few fair-weather locales in the United States such as Savannah and New Orleans. During the 1880s and 1890s, the average amount of ice harvested in just the Hudson River and Maine was right around four million tons per year.[4] The enormous magnitude of this increase in production would not have been possible without a wide variety of innovations throughout the entire harvesting process.

To implement these changes required lots of seasonal labor. However, the kind of people working on ice harvests changed over the course of the nineteenth century. The first ice harvesters were local New England farmers who had little else to do in the dead of winter. About one hundred Irish laborers employed by Tudor disturbed the peace of Henry David Thoreau while he was living on Walden Pond during the late 1840s.[5] When the size of operations grew, harvesters needed to import day laborers from far away to handle the workload. The number of workers employed just on the lower Kennebec in Maine during the 1890s usually reached around forty-five hundred.[6] Harvesters constructed boardinghouses for men who came from all over the East.[7] Along the Hudson, there were often twenty-five thousand men working to cut ice.[8]

When ice harvesters no longer recruited exclusively farmers, they began to have trouble with the reliability of their workforce. Ice companies in New York City deliberately recruited the unemployed and chartered special trains to take them to ice fields throughout the Northeast. "A great many of the men we hire . . . are unreliable," complained one Wisconsin dealer. "They work a day or two, then make some excuse for drawing their wages, and we never see them again. I don't know who these men are, but I think they are tramps."[9] Harvesting was hard work—hard enough to make any man unreliable if he wasn't used to that kind of intense labor.

One reason that ice harvesting could be so difficult was that harvesters generally tried to work as fast as possible so that they could avoid the pitfalls that a weather-dependent business brought. This seasonal work, lasting three or four weeks in most seasons, was entirely dependent on the weather. If it got too cold or too snowy, harvesters had to suspend their operations. How cold was too cold? "Cheer up, boys, it has got up to ten below," joked one superintendent of an icehouse in Maine early in the twentieth century.[10] It took a hearty person to work under such difficult conditions.

Cold, clear days were the best days for both making and harvesting ice since prolonged periods of cold led to the thickest ice. When the ice season started early thanks to extended periods of cold days without snow, harvesters had the opportunity to harvest the same ice fields multiple times. The more ice they harvested, the more ice consumers could buy. Whatever the ice companies lost through low prices due to increased supply, they could usually make up through higher volume. This was particularly true during the late nineteenth century, when the market for ice was rapidly expanding.

On the other hand, warm winters could obviously destroy a whole year's harvest. Rains during winter were an even bigger problem since they destroyed whatever ice had already formed when warm water collected on top. Even if some ice remained after it rained, that ice tended to be covered in dirt and other sediment carried onto the field with the water. Even an inch or two of water on the ice could leave it full of holes, making it unsellable. The worst thing that could happen to an ice harvest was a freshet, a flood on the body of water where they wanted to harvest. In those instances, there probably wasn't any ice that was thick enough to harvest, if there was any ice at all.

No matter what kind of worker you happened to be, ice harvesters faced a great risk of accidents. Despite special spiked horseshoes, the animals could slip on the ice and break a limb. Both men and horses fell through the ice, sometimes together. Once the huge stacks of ice in icehouses reached near the ceiling of these huge buildings, the people finishing the stacks risked falling from great heights. While cutting ice on rivers with any kind of current, workers could suddenly find themselves on an ice floe headed miles downstream with no way to return but by walking. Yet there was generally a surplus of unemployed people looking for work in the rural areas where ice was cut because they had no other options for employment at that time of the year during the late nineteenth century. By then, ice harvesting had become a tradition that attracted both job seekers and spectators.

Many observers described ice-harvesting operations during the heyday of this industry because few ordinary people understood how this process worked.[11] This was especially true later in the nineteenth century as the process grew in scale and utilized more tools than before. Nevertheless, the basic methods by which people cut natural ice during the late 1800s remained remarkably similar to the methods that Frederic Tudor had developed by trial and error earlier in the century.

Before any harvesting could occur, harvesters had to use special scrapers to remove all the snow from the ice. If the snow stayed on the ice for any length of time, it would insulate the ice under it, thereby making it melt faster. Usually scraping occurred every time snow fell, whether there was any immediate harvesting or not, because frozen snow was not a viable commodity. Hence, the size of the snowbanks at the edges of the ice fields revealed how far the

Ice harvesters on a frozen lake, ca. 1912. Library of Congress

harvest had proceeded or how many times the same bodies of water had been harvested.

A common strategy to get thicker ice was to "wet down" the first snowfall. A group of workmen, each using a ring chisel, cut holes about six feet apart across the ice field. The weight of the men on the ice bent it down and forced water up through those holes. Once on the surface, that water would get absorbed by the snow. This would serve as a barrier to protect later ice accumulation. This compressed the size of the total ice and snowpack to between one-third and one-quarter of its original thickness, thereby making it easier to cut and move.[12] Once the real harvesting began, whatever snow-ice remained could be planed off in favor of the clear blocks of uniform ice, which fetched the highest price.

Before the cutting started, teams checked to see whether the ice had reached a proper thickness of at least eight inches using a drill and an ice auger. Generally, the ideal situation was ice between twelve and fifteen inches thick without any snow on it. Whenever they began, the harvesters would line the field off into squares. Two straight lines—about twenty-two inches apart, the width of a regular block of ice—were marked by a hand plow, which would be run along the side of a long board. The space between these lines would serve as a channel to float ice blocks from where they got cut to the edge of the water, where they would eventually be removed and stored. At this point a marker, an implement with cutting teeth which looked like a low plow, would be drawn by a horse over the course of the initial straight lines. With two trips, back and forth, the marker could cut a three-inch seam into the ice. Harvesters would mark off the entire ice field before any cutting began.

While determining the depth of the field, harvesters would note the quality of the ice as well. If there were a sizable portion of snow-ice on top, they would have to scrape the ice before beginning the harvest, especially if the snowfall had made the ice uneven across the top. These scrapers were usually pulled by horses. If there was a lot of snow, the scraper the harvesters used looked more like a scoop than a planer. While planing and scooping made for better ice, removing the chips this process produced could slow everything down considerably. Still, they had to be removed since the hail-sized particles produced by the planing were a safety hazard for men and horses alike. Even if the horses avoided tripping on them, sharp shards of ice could cut their ankles to pieces.

Once cleared, harvesters ran ice plows over the seams to cut the ice further. These first appeared in 1839 and gradually improved over time, based on experience. The first ones had wooden beams. These were replaced by iron beams, which were in turn replaced by steel. By the 1880s, ice plows had completely replaced saws and axes as tools for ice harvesting.[13] The guides to make sure that the blades stayed in the grooves were what separated an ice plow from a regular plow. These allowed harvesters to run the blade back and forth along the same groove until the cut reached the exact thickness that they desired.[14] While hardly a perfect process—it could take three men to make sure that the plow hit the grooves just right—this was an essential step for producing a regularly shaped block of ice, an essential trait for storing and transporting the final product.

Only small areas of an ice field could be marked and planed at one time because invariably the ice would be at different thicknesses at different spots. The size of the ice plow needed varied depending on the thickness of the ice. Later plows had steel teeth so that they could be used to carry out the chips, or narrowing beams so that the chips could shoot out one end like a modern lawn mower.[15] Harvesters had to have many plows on hand in order to cut ice to whatever thickness they desired. Having a lot of plows also helped finish the cut, since ice tended to freeze together in the region getting plowed; how fast it froze depended on the weather. The general tendency was to cut the blocks to about two-thirds of the depth of the ice on first pass.

Once the first grooves were cut, harvesters used a caulking bar to fill in any holes in the blocks that the cutting exposed. Ice chips from earlier in the process were the usual filler. Harvesters who didn't caulk risked having the blocks that they precut freeze together again as water filled in the holes along the sides of the cuts. The risk of water getting into the holes was particularly large in a tidal river like the Kennebec in Maine, where the movement of the water could be quite strong. A moderate wind could cause the same problem just about anywhere.

Once the channel to the icehouse was finished, the harvesters would begin to detach ice sheets, usually made up of about ten blocks, from the spot on the field farthest away from the icehouse. Then they would turn those large sheets into separate blocks with a surface of about forty-four square inches. The usual way to do this was with a breaking bar, which was a long, broad, wedge-shaped chisel. Those groups of blocks would be drawn along the open water by horses until they reached the edge of the water nearest to the icehouse. There men would break the groups apart with a bar, after which they were raised and stored. During evenings when the temperature stayed below freezing, somebody had to walk up and down along the canal stirring the water with a pole; otherwise, the harvesters would find that their canal had frozen over when they returned to work the next morning.

This complicated process required a lot of tools because there were so many different steps needed to create regularly shaped ice blocks and to get them all the way from a body of water into a neighboring icehouse. The Wood Company, the largest ice tool manufacturer in the United States, sold at least two hundred different kinds of them during the mid-1880s, but this included variations of particular tools designed for the ice in different regions.[16] "Great care and much study have been devoted to perfecting the paraphernalia by which the ice crop is handled," wrote the author of an 1893 treatise on ice harvesting. "In some emergencies, the crop is only saved from partial or complete loss by the despatch which [the] modern development of tools and machinery has placed at the command of the enterprising ice dealer."[17] Of course, the cold, tired men on the ice fields used those tools, not the people who owned the ice cutting and storing enterprises.

Luckily, besides making ice harvesting more efficient, ice tools also made the work at least a little easier for the people doing the harvesting. One way to do this was to physically separate the workers from the product they were harvesting. A St. Louis reporter noted in 1879 that "all through this frigid expedition . . . at no time did workmen actually touch the blocks of ice."[18] The reporter complimented the boss for his thoughtfulness, but it seems more likely that this would have been

Workers loading ice on an ice elevator for transportation to the icehouse, ca. 1910–15. Library of Congress

a requisite for finding workmen capable of lasting through the harvest without having their fingers freeze. Indeed, most people could only stand direct contact with ice for about an hour, even when they wore gloves.

While the traditional ice-cutting tools that made it possible to avoid touching the ice were ingenious, it took other, more industrialized inventions to make a significant expansion in the volume of ice harvested possible. For example, steam-powered ice saws began as a tool for polar exploration, or to get ships out of more temperate ports after they had been frozen over. During the 1890s, smaller gasoline- and electric-powered versions came to ice harvesting so as to speed up that process. The electric-powered basin saw, which had several circular blades in a row, cut up sheets of ice before they entered icehouses. This greatly increased the efficiency of the storing process.[19]

Efficiency was also increased by improvements in ice elevators—long conveyors that lifted ice blocks out of the water and into an icehouse. By 1869, icemen in Massachusetts had started using steam-powered conveyor belts and ice elevators powered by steam rather than horse-drawn sledges. Steam sometimes powered chains that took ice blocks out of the water; they were an absolute necessity for raising blocks anywhere near the top of enormous icehouses. From there, they could be slid to any corner of the building using specially designed slides. Called "endless chains" at that time, they could lift as many as fifty blocks (or twenty-five tons) of ice per minute as early as 1883.[20] Sometimes the steam engine that powered the elevator also powered other parts of the ice-cutting process, such as saws or other elevators for loading railway cars.

Those slides that got the ice from the top of the elevator to its final resting place for the season deserve appreciation too. A 1908 article from the *Omaha Bee* described an icehouse owned by the meatpacking company Swift, with an eighty-thousand-ton capacity, as a technological marvel: "The runways are a series of platforms outside the house at different stages from the ground up, slanting all ice six inches toward the center of the house for every sixteen feet of ice. This protects the sides of the building. To glance into that large building before it is quite full is to see a rather imposing spectacle. One might imagine all the ice in the world was packed in there."[21] A large icehouse required at least twelve doors to receive blocks so as to get it to every part of a building. If the run were too steep from the top to the bottom level of ice, the ice would crack when it hit its final spot against the ice that was already there. If it weren't steep enough, the men packing it would have to work harder, and it would take more time.

Ice got packed from the back of an icehouse to the front and from the bottom of the icehouse to the top. It was best to pack ice on cold days, when the ice could be placed in the icehouse while cold and dry. In warmer weather, harvesters had to deal with more of the water that icehouses were designed to avoid. Workers used ice hooks and tongs to move the blocks into place, and bars and chisels to level off the cakes so that the next one entering the house could be packed on top of it. Ice runs and wooden skids helped icehouse workers get the blocks into their long-term positions. After placement, they would coat each block with a layer of sawdust so that they wouldn't stick to the block placed next to it. Getting as much ice as possible into an icehouse helped nobody if the ice companies couldn't get it out in warm weather. Therefore, not only packing techniques but the design of the icehouses themselves focused on keeping units from melting rather than holding the maximum volume of product.[22]

Thanks to years of trial-and-error experimentation, icehouses during the late nineteenth century could keep ice much better than the makeshift shacks that held ice during the early days of the industry. "The house into which the cakes are now hoisted is quite a different affair from the covered pit or cellar, or the hill-side cave of from sixty to eighty years ago," wrote Henry Hall, in his Bureau of the Census report on the ice industry from 1883.[23] The best ones could keep ice solid not just until summer but for several years, until the market demanded it.

Since melted ice is always warmer than ice itself, icehouse designs included a way to drain water out the bottom. The use of sawdust as insulation dated from early in the century, not just in icehouses but also when packing ice in ships going all the way to tropical climes. Demand for sawdust only increased as the natural ice industry grew in volume. Nobody really needed an icehouse to keep ice from melting. Stack enough of it outside under a tarp (something that ice companies did in particularly productive years), and there will still be a lot of it left in the summer. The secret was to preserve as much as possible so that it could make money for ice companies whenever the market would pay the highest price for it.

Most icehouses were large wood buildings made of spruce, hemlock, or white pine, erected at the edge of the riverbanks wherever the ice would be cut, so that the ice could be stored until the summer, when it would fetch the highest price. The largest icehouses on the Hudson were one hundred feet long (or more) and forty feet wide. In Maine, the icehouses could be as long as three hundred feet. Many had multiple stories, with additional ice elevators mounted on outside platforms to get blocks of ice to the very top of the house.

The outsides of icehouses were covered in sheathing or even ship-siding, painted bright white to reflect light that could heat up their interiors. The doors in commercial icehouses stretched from the floor to the ceiling and would gradually be closed as it filled up with ice. A good-sized icehouse might hold sixty thousand tons. Despite being built from heavy timber, the walls of icehouses tended to bulge after just a few years. In that case, ice companies built heavy braces against the walls so that these houses might stay in operation a few years longer.

Perhaps the most important innovation in the later icehouses was to subdivide them into rooms. Sometimes companies built separate structures under one huge roof with gaps in between, but this eventually became separate rooms in the same house. This innovation served two purposes: it allowed for the circulation of cold air through the walls in the interior of the houses, and it made it easier to get ice out of the house in batches without harming the conditions that kept the rest of the ice cold. Sometimes, especially in places where the winter air wasn't always all that cold, sawdust or hay went right into the walls to serve as further insulation.

The floors of commercial icehouses were also important to the success of the entire building. The very bottom was a layer of stones. Shavings or straw went above the stones, followed by a layer of charcoal to absorb water. Wood boards went above the charcoal. The entire floor sloped toward the center, where a trench dug from end to end could carry the melting water out of the house. Additional trenches were also built along the side of the building for the same reason.

Even under good circumstances, water adversely affects all types of wood. Therefore, a typical icehouse lasted only about fifteen years. (Stone houses, usually built for private mansions before the nineteenth century, obviously could last much longer.) Modern wood preservatives had not yet been invented, so icehouses would require a lot of off-season repairs to last even to the end of their normal lives. A primitive wood treatment called "black ile" was used in many Maine houses. Even then, many boards would have to be replaced every single year because of exposure to all that meltwater.[24]

Icehouses also faced a major threat from fire. The larger icehouses of the late nineteenth and early twentieth centuries were invariably by far the tallest structures wherever they got built. This made them susceptible to lightning strikes during thunderstorms, making lightning rods particularly important for icehouses. While icehouses were wet places and filled with ice half the year, they still had more than enough dry flammable material in them to burn easily. This fact made it very difficult for owners to insure them.

The existence of the natural ice industry in Maine was the result of that open winter on the Hudson in 1880. New York companies started an ice rush in Maine in order to meet as much of the demand as possible. While some harvesting had occurred there in previous years, it was the huge increase in demand from New Yorkers that led to a spike in harvesting there. "There is plenty of ice in Maine," one ice company executive told the *New York Tribune*. "The New York companies are chartering all the vessels they can get for the season."[25] By the end of that summer, 1,300,000 tons of Maine ice had been cut, and it got dispersed not just to New York but also to cities as far away as Charleston, South Carolina.[26]

While there had been some ice harvesting in Maine before 1880, this was the first season in which huge vessels sailed up the Kennebec River and shipped frozen water south by the ton. An acre of water might yield one thousand tons of ice each season.[27] It helped that Maine ice—particularly Kennebec ice—proved to be of particularly high quality. The Kennebec was a tidal river, which meant that the water in it moved back and forth rapidly. This, in turn, meant that the ice produced from it contained very little sediment.

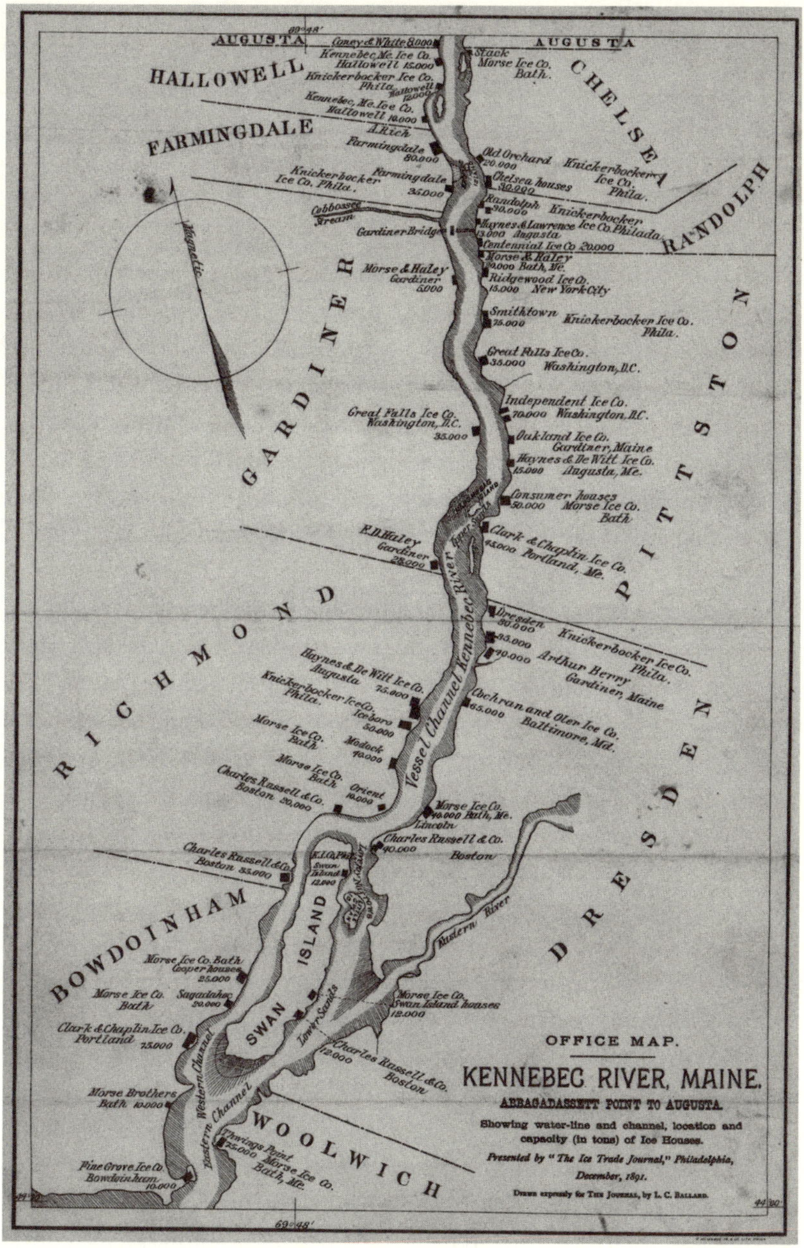

Map of ice-harvesting locations along the Kennebec River, by L. C. Ballard for the *Ice Trade Journal*, Philadelphia, December 1891. Collections of Maine Historical Society, Map FF-81

The Kennebec was also particularly clear and deep, which meant that the ice produced from it was translucent and blue. "Clear and blue as the sky," was a slogan that appeared on many ice wagons advertising Kennebec ice in cities all along the East Coast.[28] One of the reasons that the Kennebec was such a successful place for ice harvesting was the fact that ships could travel all the way up this particularly deep Maine river (either by themselves or sometimes with the help of a tugboat), right to the point where the ice got harvested.

Because of the quality of Maine ice, New York firms and others in cities farther south along the Eastern Seaboard bought tons of ice from there even after the Hudson River ice crop returned to normal. From 1,300,000 tons during the ice rush of 1880, the total output of ice from Maine grew steadily to a little less than 3,100,000 in 1890. About half of that came from the Kennebec River alone each season.[29] Since the Kennebec was wide enough and deep enough that large barges could come right up the river, the ice didn't need to be packed onto trains before getting loaded on ships, as it had been around Boston earlier in the century.

The quality of Maine ice demonstrates the general differences in ice from various locations. Ice from lakes fed by springs was the best. Ice cut from clean running water, like the Kennebec, was equally good. The air and the current helped clear the water before it froze. As the century progressed, water sources around major American cities got much more polluted. This explains why people were willing to go as far afield as Maine to get ice. There were even differences within the ice cut from a single source. Since ice grows from the bottom, the clearest ice would always be cut from there, while the top of even the clearest ice was likely to have at least some sediment that blew onto it. If that ice got dropped into people's drinks, the sediment would end up on the bottom of people's glasses. So while the poor settled for local ice of inferior quality (and paid dearly for it), the market for clear, clean ice from the Kennebec grew considerably over time.

It helped that 1890 was another open winter—a winter for ice harvesting that proved even worse than ten years before. The Hudson River failed to produce a single ton, and the weather stayed comparatively warm all season as far north as Boston. Many New York icemen went to look in Upstate New York for ice, but there was no infrastructure for getting ice out of upstate lakes and ponds since it was seldom cut from those regions before that year. Even cutting on the northern reaches of the Hudson proved difficult, since the large barges that took ice from the lower Hudson into the city could not travel that far north because the river was too shallow for them.[30] In Maine, on the other hand, additional barges could fit easily. The total harvest in Maine that year exceeded the amount taken during every other previous year by over a million and a half tons.[31]

From Maine, as it did ten years earlier, the ice would travel to large cities all along the Eastern Seaboard. The Knickerbocker Ice Company of New York took much of the cargo, as did large ice companies in Philadelphia, Baltimore, and Washington, DC.[32] The round-trip voyage between Washington and Maine took six weeks. One vessel might make that trip twice each season. Only 5 percent of that ice would melt on the trip.[33]

Another famous source for high-quality natural ice were the lakes of Wisconsin. Many of them had been created by dams, which made them both deep and exceptionally clear. Because of the weather and the water movement on most of these bodies, the ice created was usually between twelve and fourteen inches thick—in other words, perfect for packing. Earlier in the century, much of the ice cut in Wisconsin went for brewers in Milwaukee. Pabst, for instance, did not switch over to ice machines until 1888. Other Wisconsin brewers made the same transition later. Chicago meatpackers were the other major buyers of Wisconsin ice, and they made the transition to mechanical refrigeration at about the same time. Armour, for instance, did so in 1888.[34]

New York City was the chief consumer market for natural ice in the United States as early as 1880. By 1879, New York and the then-separate city of Brooklyn consumed 1.3 million tons of the stuff. This constituted more than one-quarter of all the ice consumed in the United States.[35] A key driver was the huge market for ice among the poor. When New York ice dealers hiked prices by 50 percent during the summer of 1889, the *New York World* complained, "Unless something is done in the matter there will be widespread suffering among the poor people and a higher death rate than ever among the sick babies of the tenements during the hot weather." The paper also noted that ice had become an "absolute necessity" during the summer.[36] Therefore, complaints against ice monopolists appealed to people of all classes.

Politicians eventually acted as a result of such indignation. During the great heat wave of August 1896, the city of New York, under the direction of Police Commissioner Theodore Roosevelt, bought up tons of ice and distributed it for free to the poor. "It was to them better than bread to the starving," wrote a reporter for the *New York Journal*. "Mothers with sickly babes in their arms jostled with weary men, while children begged for a piece of ice to take home to the sick room."[37] Rich people could afford clear, clean ice from Maine, but poor people bought whatever ice they could get their hands on. In bad years, that ice might come from just about anywhere.

These open winters directly promoted the consolidation of New York's natural ice industry. Charles W. Morse, whose family made its fortune in the Bath, Maine, shipbuilding industry, began to invest in the Maine ice industry during the 1870s, while he was still in college. He bought ice from local operations and shipped it to New York City in his own ships. He made $500,000 before he had even graduated. He settled in New York City in 1880 and watched his money grow as the Maine ice industry expanded.

After the ice famine of 1890, Morse began to buy New York City ice companies. He merged two different New York companies into the Consolidated Ice Company in 1894. In 1896, the owner of the Knickerbocker Ice Company in New York sold out to Morse. In 1899, Morse's firm merged with all the great northeastern distributors to form the American Ice Company, better known as the "Ice Trust." The stock was, to use a term from that day, heavily watered, which meant that the value of the stock sold to investors greatly exceeded the real value of the firm. While the previous companies had stock worth $20 million, the new firm was capitalized at $35 million. Morse tried to make up the difference by raising the price of ice on unsuspecting New Yorkers.

In the summer of 1900, right after the merger, Morse doubled what he charged for a ton of ice. That price increase quickly made it down to the poor, making ice unaffordable for many of them. "ONE HUNDRED PER CENT RISE IN ICE" read a six-column headline in the *New York Times* from May 6 of that year. "New York's Trust Limits Harvest and Controls Distribution—American Company, However, Has a Monopoly and Squeezes Rich and Poor Alike."[38] Morse's hometown paper back in Maine claimed that the high price of ice in New York City increased the city's death rate by 5 percent.[39]

Morse's political connections are what made this kind of bonanza possible. A lawsuit by the *New York Journal* led to the shareholders of the American ice company being exposed. They included Mayor Robert Van Wyck and thirty other New York City politicians, judges, and officials. The stock received by the mayor and his brother was worth $1.7 million. As a result of these revelations, the stock of the company started to decline rapidly, and the firm eventually had to be reorganized. The *World's Work* later reported that Morse nevertheless made $12 million for his troubles.[40]

Unlike many other trusts from this era, the controversy over Morse's arrangements led to criminal and civil investigations of the American Ice Company and numerous hearings in the New York State Legislature. While this kind of publicity ruined Morse's reputation, nobody ever got charged with anything. As the muckraking journalist Gustavus Myers later explained, accusations against Morse "were not based on any concern for the mass of people, nor upon any indignation at the privations, sufferings and deaths caused by the methods of the Ice Trust; they were made solely on behalf of the smaller firms whom Morse had forced out of business."[41] However, outrage over the controversy did lead to the ouster of Mayor Van Wyck in favor of a reform ticket in 1901. Morse quit the ice industry in response to all this controversy, allowing him to concentrate on other interests, using his ice profits to fund these efforts. Ironically, Morse would be imprisoned at the federal penitentiary in Atlanta for unrelated bank fraud in 1908.[42]

The rise of Charles Morse demonstrates the importance of numerous barriers to entry in the natural ice business. Once someone had created the infrastructure needed to harvest, store, and distribute ice, it became very difficult for others to compete. New players in that field would have had to pay high up-front costs to duplicate those functions, and if they passed those costs onto their customers, they would not have been able to get a toehold in the market. Regional ice monopolies could be very profitable since ice had become a necessity for consumers of all classes by the turn of the twentieth century.

Because ice production and distribution were so decentralized, there was not one but many ice trusts in the United States around the turn of the century. They were particularly strong in Chicago and St. Louis, but they were unrelated to Morse's company. As the prosecutor in Toledo, Ohio, home of another strong ice trust, put it, "In most cases a local trust exists, but it is impossible now to find evidence that the trusts of various towns are related to each other however much their relationship may be suspected."[43] Indeed, prosecutions against the owners of ice companies for state antitrust statutes were a fairly common occurrence during the early twentieth century. However, Morse attracted the most attention in the national press because he had gained control of the largest market for ice in the country.

Morse built a far-flung vertically integrated company so that he could charge whatever he wanted for the ice that he harvested. Even if Morse's prices attracted competitors in Maine, they would have had difficulty transporting all that ice southward and then distributing it around the city because it took decades for this particular cold chain to emerge. Had a competitor decided to harvest ice near New York City, he would have had trouble finding clean water to harvest ice on because the Ice Trust already controlled the entire supply. Technological change, not the law, was what eventually destroyed the Ice Trust and its brethren.

The improvements that natural ice harvesters made later in the century protected them from the competition they would face as the machine-made article gradually improved. The number and effectiveness of ice tools grew over time. Since these ice tools and especially the practices of natural ice harvesters couldn't be patented, they spread throughout the northern parts of the country to wherever ice happened to be harvested. Nature determined whether that year's crop of ice would be large or small, but ice harvesters employed countless different technologies in order to ensure that they could cut and preserve as much ice as possible until the summer, when customers were willing to pay the most for it.

Ice harvesting may seem primitive compared to other industries of the era like steel making or automobile manufacturing, but it too was at least partially industrialized. While not fully automated, natural ice harvesting benefitted from the residual developments associated with industrialization, such as electric power. Over time, the price of labor dropped as machines like ice elevators replaced men for at least some of these functions, but not enough to save the industry from its eventual demise.

The natural ice industry still faced constant pressure on price even as the efficiency of its harvesting methods continually improved. Customers complained constantly about costs, and to some extent they had a point. Since natural ice cost nothing to manufacture, harvesters were essentially in the ice-handling business and sometimes also the delivery business. However, their customers seldom recognized how much it cost to store and move ice. This explains the ice harvesters' common practice of never talking about how much ice they kept in their houses. If they had a lot, people would expect lower prices even when they had little ice on hand. If they had little ice on hand, customers would generally complain anyway.

Finding complaints about ice prices in the news coverage of the ice industry during the late nineteenth and early twentieth centuries is child's play for anyone who looks for that evidence. What is hard to find are customers willing to swear off ice entirely. Yet at least some consumers tried. "I have been taking five pounds of ice every other day," one customer told a reporter from the *Chicago Daily Inter Ocean* in 1896, "but yesterday, having received notice that the price was advanced 10 cents a week, I turned my back on the iceman and canceled my order. I can do without ice for a month, anyway, and I do not purpose being held up so early in the season."[44] Even in a monopolized market, customers like this one could choose to do without. The more competition that existed, the more volatile the price became. That seldom worked to the harvesters' benefit.

Faced with complaints about price and periodically inspired to look for an alternative by periodic open winters, entrepreneurs of all kinds experimented with ways to create cold without the need to harvest ice as early as the 1860s. During the height of the ice famine in 1890, the journal *Scientific American* received fourteen letters in a week asking for information about mechanical ice machines.[45] Everyone looked forward to the day when ice machines would become efficient and reliable enough to produce enough ice to satisfy the demand. Many observers predicted that this would happen quickly. In fact, it took another twenty-five years or so for that prediction to become true.

The natural ice industry persisted as long as it did because it suited segments of the ice market particularly well, especially poor people and people who lived close to the bodies of water where ice got cut. It may have been dirty. It may have melted quickly. However, ice was a very cost-sensitive product for most consumers, so natural ice lasted as long as it stayed cheap.[46] Various problems with the development of mechanical refrigeration also helped keep natural ice as a viable product until the World War I era. Even a solid block of ice full of sediment collected from a lake or stream was better than a poorly made block of artificial ice that would melt quickly. From inception to perfection, the development of a dominant artificial ice industry took almost seventy years.

2 How to Manufacture Ice

DURING THE 1840S, an Apalachicola, Florida, physician named John Gorrie wanted to treat the victims of malaria there. Recognizing that the disease disappeared during cold weather, he invented a machine to bring cold weather on artificially through mechanical refrigeration—what today we would call air-conditioning. His apparatus depended on suspending a basin full of ice near the ceiling of a room and using a fan to blow air over it so that that air could be purified. He thought that nobody could catch malaria in such an environment. Unfortunately, the cost of natural ice in the South was prohibitively expensive at that time.

Needing a more reliable source of ice than the natural ice from New England which sporadically appeared in local ports, Gorrie quit his medical practice and set out to build an effective artificial ice machine. He filed for his first patent in 1849 and demonstrated its operation publicly in 1850. While the design of the machine worked in theory, creating a workable ice machine in practice proved much harder. Gorrie constructed his original machine in Cincinnati, but because of poor local craftsmanship, he could not get it rebuilt so that it worked right after he shipped the parts to New Orleans. He could not even get a thermometer made in New Orleans which would give him a reliable reading. The resulting machine leaked refrigerant and performed irregularly.[1]

Obviously, such problems led to the commercial failure of Gorrie's efforts. No evidence exists that anything beyond his prototypes were ever licensed or produced. These problems probably hastened Gorrie's death in 1855 at the age of 51. Yet even had Gorrie managed to manufacture a reliable ice machine, he faced the problem of making a profit. As Gorrie himself explained shortly before his death, mechanical refrigeration "had been found in advance of the wants of the country."[2] In other words, even in sweltering New Orleans, the market for artificial ice could not support all the costs associated with Gorrie's invention.

America did not have a particularly strong taste for ice in the early 1850s, when Gorrie went searching for investors.[3] But it did thirty years later. This was a sign of the great impact that the ice industry had on the American diet. Previously, people were content to keep pickling, canning, and storing food in root cellars in order to preserve it. Despite all this work, ice was still too expensive for the great majority of households to use regularly, especially during seasons when there was any kind of ice shortage, so they used these other food preservation methods instead. This became less of a problem when artificial ice eliminated the reliance of the ice industry on the weather, but even then it took years for ice machines to become a reliable enough technology to justify their large initial costs. In the meantime, the ice harvest continued.

Gorrie's failure is an excellent reminder of the role that markets play in the success of new manufacturing technologies. Even the best new technologies won't displace old ones overnight. The products they make possible require markets and consumers who have enough assets that they can afford to buy those products. Lack of these things explains why it took forty years for mechanical refrigeration to completely displace nature as the primary way to produce ice. In part this was a testament to just how efficient ice harvesting had become. Even when ice harvesting finally disappeared, mechanical ice-making facilities significantly shortened the cold chain for providing ice, leaving more to sell.

Costs were another problem that explains the lag between the ice industry and the mechanical refrigeration systems that eventually replaced it. Just because somebody invents something useful does not mean that they can necessarily make a profit producing or using it. Their costs mattered. The increased efficiency of all the technologies used along the cold chain, as well as years and years of marketing by ice companies, eventually impacted the American public's willingness to make ice part of their regular diets.

Machine-made ice gradually became the norm in the United States thanks to many small improvements in efficiency. In Gorrie's day, natural ice was cheap enough in northern climes that it could more than satisfy that underdeveloped market. When mechanical refrigeration became cheap and convenient enough that it could compete against natural ice even in the far northern parts of the United States, the market for natural ice gradually shrank, until it eventually disappeared entirely. Shortly after natural ice ceased to be economically viable, the first large ice machines were eclipsed by electric household refrigerators, and the ice market changed once again.

It may seem strange to closely consider the history of large, commercial refrigerating and ice-making machines that only lasted a few decades, but the advent of refrigerating machinery that made tons of ice each day during the late nineteenth and early twentieth centuries was a milestone in human history. Before mechanical refrigeration, human beings were completely dependent on nature for controlling the cold. In places that never got particularly cold, obtaining ice in significant quantities was essentially impossible. With large quantities of manufactured ice, it became possible to defy seasons for the first time. This technology also provided the first way anyone, anywhere could really stay cool in summer and the first way to preserve perishable foods so that they could be eaten out of season in temperate climes. These developments were life changing at the time, and even though ice plants are no longer with us, the effects that our initial forays into mechanical refrigeration brought persist to this day.

Local markets that had once depended on New England ice first began to experiment with manufactured ice during the late 1860s. Most of them were in port cities in the American South. Around the same time, ice machines provided the first ice to southern cities located further inland. These machines came about as the result of engineers experimenting with refrigeration equipment developed in Europe. Unfortunately, operating costs often proved extremely high. Perhaps most importantly, the ice that these machines produced was thin and therefore melted quickly. Given a choice between artificial and natural ice, consumers would invariably pick natural ice until artificial ice could produce a better product. The problem with the ice was the result of design flaws in these early machines.

Many different inventors from around the world experimented with ways to create mechanical refrigeration during the earliest years of this technology. The result was the compression refrigeration cycle.[4] Under these systems, various liquids could be heated, and as they changed states, they could draw heat away from adjoining spaces. The first successful ice machines were huge, factory-sized contraptions that could weigh as much as five tons. People who built these large machines invariably charged more for their ice in order to earn back their costs.

Because artificial ice was expensive, few people wanted to operate an artificial ice machine in places where there was an abundance of cheap natural ice around.

The first ice machines appeared in America during the 1860s; however, the first really successful ice machines did not appear in America until the 1880s. For example, the Frick Company of Waynesboro, Pennsylvania, a firm previously known for making agricultural implements, steam engines, and boilers, introduced its first mechanical refrigeration machine in 1883. It was specifically designed to manufacture ice. In 1888, the firm promoted its "Eclipse" ice-making machines by comparing the advantages of artificial ice with the practice of harvesting ice from rivers, lakes, and ponds. "Your temperatures do not vary with the weather," it wrote in its sales pamphlet. "You are not dependent upon the fluctuating price and quality of ice. If it is an open winter and no ice crop, you are independent, your business goes right on during the warm weather." Read the entire pamphlet and you can see many examples of the firm's considerable bravado. "The ice dealers do not like Refrigerating machines!" it noted elsewhere in its same catalog. "Why? Because it hurts their business; it is a money-saving institution and they can't compete with them; that tells the whole story in a nut shell."[5] Actually, the natural ice industry would compete successfully against the machines made by firms like the Frick Company for another twenty years.

A serious problem facing ice machines was the potentially explosive ammonia that most facilities used as the refrigerant. "There is much astonishment expressed to-day that such a harmless and innocent-looking piece of" machinery as an ice machine, explained the *New York World* after one such explosion in a packinghouse, "should carry with it such disastrously fatal results."[6] Such stories demonstrate why the public was afraid to have ice machines operating in their neighborhoods well into the 1920s. They blamed any accident or fire in a refrigerating plant on explosive ammonia gas, whether the refrigerant was at fault or not.[7]

The primary problem in 1880, however, was still the poor quality of the ice that machines could produce. A *Scientific American* article entitled "The Problem of the Ice Supply for New York City" observed that even though companies in that metropolis manufactured ice with machines, none of those machines was in use in New York City at that time. Those machines were all sold farther south, where the alternative to poor-quality artificial ice was no ice at all. "In its present state of development," that journal noted, "ice making machinery is not able to compete with Jack Frost in our climate."[8] In other words, natural ice was so much cheaper than ice machines that the only market for the artificial product was in places where the temperature never dropped below freezing. Over time, however, gradual improvements in artificial ice-making technology made artificial ice better and cheaper, and improvements in the machinery made it more reliable.

That explains why many people imagined a great future for mechanical refrigeration, especially during the 1890s. For example, a reporter for the *Omaha Bee* observed in 1897 that "the outlook now is that, with the introduction of a process just perfected, the cost of making artificial ice will be cut down to such a figure that natural ice cannot compete with it, even in such northerly cities as Montreal or St. Paul."[9] While the mechanical refrigeration process had not yet been perfected, this prediction did come true during the 1920s. The delay could be attributed to the fact that even the most skilled engineers of the era had little understanding of how an ice machine actually worked.

No matter how an ice machine produced ice, each one created cold through the same basic principle, which had been developed from the research of a series of inventors working around the world. The compression system remains the basic framework by which refrigeration works to this day. In these older machines, however, the parts were prone to breakdowns and generally much less efficient. Over time, the source of power for refrigerating machinery changed from steam to electricity. Nevertheless, the following broad description of the compression cycle describes both the inner workings of these old ice machines and the process behind modern mechanical refrigeration.

The compression refrigeration cycle depends on the compressor forcing a refrigerant around a system of coils. A refrigerant is any substance that can be used to draw heat away from an adjoining space, but some refrigerants worked much better than others. During the late nineteenth century, most American refrigerating machines used ammonia as their refrigerant. The main advantage of ammonia was that it was very efficient. In other words, it has a very low vaporizing temperature (or boiling point) at which it will turn from a liquid into a gas. This means that it required less energy to propel it through the cycle and remove heat from whatever space or substance that the operator needed to become cold. If ammonia leaked through the pipes of these early machines (which it was prone to do), under certain circumstances it could even explode, as the New York packinghouse example described above illustrates.[10] Most American refrigerating equipment manufacturers didn't realize that until ammonia compression refrigeration systems had become extremely popular.[11]

Like all mechanical refrigeration systems, ice machines of the late nineteenth century had four different components. First were the coils, where the refrigerant would absorb heat from the surrounding water. The larger the area of direct contact between the coils and the area where the temperature was supposed to go down, the more heat could be transferred. Second came the compressor, which turned the gaseous refrigerant into a liquid and provided the power through the change in pressure that propelled the refrigerant throughout the machine. Third was the condenser, where the refrigerant cooled and liquefied. The last component of an ice or refrigeration machine was the expansion valve that regulated the readmission of the liquid into the pipes, where the process would begin again.

A compressor is a cylinder with a piston inside of it. When the compressor filled with gas, the piston would contract. The expansion of the piston eliminated the gas from the compressor and propelled it through the system. Compression began when the change in pressure that the compressor created forced the gaseous form of the refrigerant into a liquid. After the refrigerant got carried through an expansion valve, the pressure dropped. When that happened, the refrigerant absorbed heat from the adjoining space.

If the adjoining space was full of water, the water became ice. If the adjoining space was open, that space could be used for cold storage. Cold storage is what made it possible to keep perishable food fresh so that it could be sold out of season. The same basic technology would eventually be shrunk and used for refrigerators and air-conditioning systems. When the refrigerant left the coils, it went back into the compressor, where it got turned back into a liquid. Since refrigerants cost money, this process was always closed. This meant that the same refrigerant got cycled through the machine over and over.

In the early years, a steam engine powered the entire system. Fuel was always the most expensive part of ice manufacturing. "The manufacture of ice is peculiar in that practically the only materials which affect the cost are those which do not enter into the product, but are used in the generation of cold necessary for the production of ice," concluded a report from the Census Bureau in 1900. In other words, power cost a lot more than water or labor. Ammonia constituted less than 10 percent of the total cost of manufactured ice that year.[12] Over time, electricity replaced steam. Since electric generators were more efficient than steam, artificial ice got even cheaper as a result of this switch.

The people who designed and manufactured ice-making machinery were not the same people who operated them. The first refrigerating machine models built in America were based on designs from other countries and used refrigerants that did not prove to be commercially viable. Many of the first ice machine builders went bankrupt because of their experiments. Therefore, anyone operating these huge, complicated machines had to employ a refrigerating engineer in order to fix and run it. But early refrigerating engineers had no real training in how artificial ice machines worked. They were only trained in understanding machines in general.[13] They could only gain the experience they needed to gain expertise by working with one of these machines for a long period of time, which explains why many of the early ice-making operations did not work very well.

The high demand for ice machines in the South guaranteed that even ineffective contraptions had a market because even bad ice was better than no ice at all (or expensive ice imported from New England). In 1916, a longtime Birmingham, Alabama, ice manufacturer remembered the troubles he faced with his first machine around 1880. "When the machine started our troubles began," he told the trade journal *Ice and Refrigeration*, looking back on those days. "It made ice, all right, *but such ice!* Not being lasting, it could not be utilized for tomb stones. Being made from hard limestone water, over-ripe hen fruit was its only competitor in smell and taste." Despite such problems, these early machines were still good enough that natural ice gradually ceased to be a viable commodity, as northern ice harvesters turned to closer markets rather than ship their product to southern ports. Around 1883, the entire South began buying ice machines. Constant experimentation by the manufacturers of this equipment, as well as by the operators themselves, led to marginal improvements.[14]

But as late as 1890, substantial room for improvement remained. For example, a common problem with early ice machines was that red sediment ended up in the core of the blocks of ice they produced. Sometimes the cause of this was the use of bad water. Sometimes the problem was rust in the pipes that brought water into the machine.[15] This gave rise to rumors (often spread by natural ice producers) that consuming artificial ice was unhealthy. "It is a common belief that chemicals are put in the water that is frozen," explained *Frank Leslie's Illustrated Weekly*, "in other words, that ice artificially made is not a pure and wholesome product."[16] Ironically, artificial ice was far purer than most natural ice because of pollution in the water near most major American cities.

In direct contrast, the great advantage that even the earliest large artificial ice machines had over natural ice was that the ice they created did not have any sediment in it. Even the best natural ice would leave dirt, leaves, and the like at the bottom of the glass because these things were always in the water from which it formed. Clean, clear water created clear, clean ice. On the other hand, "It has

Men posed behind a block of manufactured ice in order to illustrate its transparency and therefore its quality, n.d. Nickerson and Collins Refrigeration Photographs, Archives Center, National Museum of American History, Smithsonian Institution

come to be firmly established that sewage-polluted water should not be used for drinking purposes, either in its natural state or in its condition as ice," reported the *New York Tribune* in 1903, but "not one New Yorker in ten thousand knows where the natural ice he consumes so greedily in hot weather comes from."[17] Many cities banned the harvesting of polluted ice, even if it was not intended for human consumption, since it was simply too easy to switch bad ice for the good stuff.[18] Artificial ice faced no such problem.

That explains why producers made extraordinary efforts to teach consumers that all ice was not the same—even different kinds of natural ice. "Pure Spring Water Ice from Ballston Lake," read one advertisement from 1896; "No sewage."[19] While this might have worked to the benefit of the Ballston Spa Ice Company (which placed the ad), it worked to the detriment of the natural ice industry overall because it reminded buyers that some natural ice really did come from polluted waters. Good water produced good ice, but what could the natural ice industry do when most of the good water disappeared?

Yet even ice from unpolluted waters could still be dirty. The ice cut from near the top of any body of water was inevitably filled with all the things that blow around outside. As one exposé on the natural ice industry noted, "Difficult though it is to bring a direct charge of typhoid infection against these sources, there is a perfectly reasonable probability that many cases of intestinal diseases have originated in such dirty masses."[20] Artificial ice presented no such risks. Not only did it generally appear cleaner than ice cut from nature, but it also had the advantage of being consistent. Ice taken from the top of a block of artificial ice was of the same quality as ice that came from the bottom.

By the turn of the twentieth century, ice machines were becoming efficient and effective enough to compete directly with natural ice for the first time. The key was automation. A reporter for the *Fort Worth Star Telegraph* explained one such machine this way: "In the interior of the drum the [refrigerant] is introduced by a small orifice under pressure. The result is that it expands rapidly, covering the entire interior of the drum and making a large freezing surface. On the exterior of the drum the ice forms very rapidly, the freezing going on at a rate of a quarter of an inch a minute. But it is not allowed to collect, for almost as soon as it forms it is scraped off by a set of knives which work back and forth."[21] When the machines could do more of the work that men once did, ice appeared faster and the cost of making it went substantially down.

Another reason good artificial ice was better than the natural product involved its grain. Natural ice came in uneven chunks. It might have no grain to make cutting easier, and it could sometimes be extremely brittle. "A few years ago artificially produced ice was soft and very perishable," explained a US Bureau of the Census report from 1908, "but that made according to the improved methods in use at present is as hard and lasting as any that can be obtained from natural sources."[22] In other words, improvements in artificial ice quality benefited ice manufacturers just as much as they benefited consumers.

Manufacturers agitated the water used to make ice while being frozen because consumers thought that clear ice was purer (even though this was not necessarily the case). Opaque ice has air bubbles in it, but stirring the water used to make artificial ice guaranteed clarity. Whether ice was clear had nothing to do with whether it was pure. Clear ice could be made with impure water. Cloudy ice could be made with pure water. But the public wanted clear ice because they could judge its appearance better than they could judge the actual purity of the ice itself. This preference dated back to the early days of the natural ice industry, when ice from the clearest bodies of water fetched the highest prices.

Early ice machines used purified water from the exhaust of the steam engines powering the entire operation. That water would be filtered to remove whatever oil particles remained in it, and then distilled so that it became fit for human consumption. In cities without good public water supplies, putting the water through this process made it much clearer than it would have been otherwise. This explains why so many southern ice companies were located near rivers, since waterways in that region were rarely as polluted as they were in the North.[23]

New ice machines from around the turn of the twentieth century used what was known as the "can ice" system. Water was frozen while in cans, whether by refrigerant pumped through the can or by submerging that can in brine. Unfortunately, these systems did not produce ice in large blocks, which were easier to sell to industrial customers. Manufacturers also faced the problem of ice cracking right in the can since some parts of the container grew colder than others.[24] On the other hand, can ice also tended to have a soft core at the middle of the block.[25]

These problems explain why later machines, developed around 1910, produced ice not in cans, but with plates. Plate ice came from large tanks filled with water that got consistently agitated by air jets. The refrigerating pipes ran through large iron plates, which would become cold when the machine operated. When those plates got inserted into the tanks, ice would form on them. Automatic steam cutters would break the ice that formed on the plates into blocks of the desired size. This ice was particularly hard and clear, like the ice that forms on a moving stream, since the conditions were similar.[26]

Clear, clean, high-quality ice went for human consumption. However, lower-quality natural ice remained a viable commodity for many years because it had a market among the poor and in industry. If the purpose of ice was to keep other things cold and that ice never came in contact with the foodstuffs it preserved, then it did not matter that the ice had sediment in it. All that was required was the cold. Grocers tended to go for cheap natural ice for this reason. So did meatpackers, since the railway cars that transported meat from Chicago to the eastern parts of the country were designed to separate the ice and the meat. Even household iceboxes had separate compartments for the ice and the food.

Because early mechanical refrigeration equipment could efficiently produce large blocks of ice, it had a huge effect on one of America's most popular beverages.

General view of the compressors and pipes, Champion Ice Manufacturing and Cold Storage Company, Covington, Kentucky. While this picture is from 1986, the machinery is much older, with the company itself dating back to 1884. Library of Congress, Historic American Buildings Survey

Brewers constituted the first big market for mechanical refrigeration because they wanted to manufacture ice, which they needed to make lager beer outside of wintertime. Practically unknown at the beginning of the nineteenth century, lager was introduced by German immigrants when they first arrived in America during the mid-nineteenth century. While popular because of its light taste and the fact that it was less filling than other beers, lager required cold brewing and storage. Temperatures during fermentation had to be kept between forty-seven and fifty-five degrees Fahrenheit. In storage, the beer had to be kept just above freezing.[27] Kept under these conditions, the resulting beer became clear and mellow while it fermented.

For a long time, this meant that lager could only be manufactured during cold months. In fact, the German word *lager* comes from the German expression "to lay up," which signifies how important the storage process is to creation of this beer. At first, American brewers tried to cut out this process by producing inferior substitutes that allowed taking shortcuts to produce a brew that seemed lager-like in appearance. The advent of the ice industry began to change that. Brewers who had access to large quantities of natural ice could brew lager during most of the year and store their beer in vast subterranean caverns.[28] Ice also made it possible to ship lager longer distances, keeping the barrels cold along the journey.

The change from darker to lighter beer which ice made possible also had an effect on the price. Lighter beer has a lower content of hops, and before the early 1880s American brewers depended on expensive, imported hops to make their product. Brewing year-round also increased the supply of beer available. In other words, expanded refrigeration through ice meant more beer, which lowered the price, allowing more people to be able to afford it.[29] This change had a permanent effect on US drinking habits, as lager beer remains the most popular beer that Americans drink to this day.

Over time, breweries became the largest market for the entire ice industry. In 1864, brewers were spending $1 million per year on ice. By 1877, the largest brewers of lager beer bought twenty to thirty thousand tons of the stuff each year.[30] Not even the total amount of ice used for the transportation and preservation of foodstuffs rivaled these numbers during the years before 1880. In some cities, demand from breweries alone was enough to cause a spike in the price of natural ice. No wonder the brewing industry pioneered the use of mechanical refrigeration.

The first successful mechanical refrigeration system was developed in Germany. It was funded by and sold to brewers. The same thing was true in America to some extent. John C. De La Vergne, one of the most successful refrigerating machine manufacturers, started his professional life as a brewer. He built his first machine for his Hermann Brewery in New York City in 1878 and continued to perfect his designs thereafter. He eventually decided to switch over entirely into the refrigerating equipment manufacturing business because it had proven so profitable.[31] His machines were generally considered the best ones on the market for at least the next twenty years. By then, of course, De La Vergne's company sold refrigerating equipment to a lot more businesses besides breweries. Nonetheless, of the 569 refrigerator plants listed in an 1895 De La Vergne advertisement, 369 of them were installed in breweries.[32]

Refrigeration by ice alone simply could not have kept up with the demand for lager in America. For example, it was impossible to produce lager in the South before mechanical refrigeration since the natural ice industry did not exist there. Moreover, vast icehouses owned by the breweries that could make lager took up space and cost money. Brewers that made lager beer were also subject to the common weather-related price fluctuations of natural ice. To make matters worse, the quality of lager beer produced with ice varied because ice did not produce a steady temperature that brewers could control. It was also difficult for brewers using ice to get the lager as cold as they wanted, since the cold produced by ice can be well above freezing.

Breweries took a chance on the unproven technology of mechanical refrigeration because they stood to gain so much from it. Having access to a reliable cold year-round made it possible to produce what had once been a seasonal brew any time of year. Another advantage was labor savings, because there was no longer any ice to carry around (not to mention the eradication of icehouses). "Now it may be said," explained *Scientific American* in 1890, that thanks to the "powerful aid" of these machines "the manufacture of beer is unlimited."[33] Indeed, before the advent of the ice machine, southern beer drinkers had to get their beer from up north. Now southerners could produce beer themselves.[34]

The problems with artificial ice machines which brewers faced in the late 1870s and early 1880s gradually disappeared as the technology improved. "I am happy to say [the machine that you have erected for me] does away with natural ice for any purpose whatsoever in my establishment," wrote one Baltimore brewer to the Frick Company in 1886.[35] His machine was effective enough at keeping low temperatures that he didn't even have to run it night and day to achieve the desired conditions for brewing lager. Because of the obvious benefits that mechanical refrigeration offered them, brewers were just about the only people who bought ice machines outside the South until 1885. After that, however, ice machines began to take over the function of natural ice in many different facets of the refrigeration business.

As mechanical refrigeration equipment improved, machines that produced ice grew larger. "The larger the plant, the lower the cost of production," explained the *San Francisco Call* in 1890. "The lower the price of ice, the greater the consumption."[36] This meant that more consumers, including relatively poor people, could now afford to drop ice in their drinks. However, it also meant that companies besides breweries could use ice for other purposes besides retail sales. The increased accessibility of ice meant that more businesses could use ice to preserve perishable foods. In this way, ice indirectly transformed the American diet.

Besides being the largest consumer of natural ice in America at the turn of the twentieth century, New York City also produced and consumed the most manufactured ice in the United States. Different kinds of ice served different markets. Natural ice from the Hudson largely went to the household and bar market. On the other hand, small producers like hotels began buying small machines as soon as the 1890 ice famine demonstrated the need to secure a more steady supply of this immensely popular article. Starting around 1893, smaller ice machines began to spread to industrial facilities like packinghouses and creameries. Over time, smaller ice machines proved cheaper and much more reliable than the huge machines built for brewers and commercial ice manufacturers.[37] This allowed those places to produce their own ice independently. Unfortunately, these initial machines required many repairs, and the business took years to become profitable.

Fancy hotels, dairies, and large restaurants had a high demand for ice—high enough that it made sense for them to make it themselves thanks to these smaller machines and sell off whatever excess they had to consumers. They could not, however, produce enough to make a dent in the price of ice all by themselves. In 1901, there were thirty ice-making plants in New York City, making a daily output of between sixty and six hundred tons each.[38] So despite the rapid growth of the manufactured ice industry, it had not grown enough to satisfy the then-considerable demand for ice in America's largest ice market.

Even by the turn of the century, competing against the Ice Trust did not seem like a winning proposition. Bigger machines owned by bigger companies could easily outproduce smaller machines owned by smaller companies and therefore undersell them on the open market. Larger ice plants appeared in smaller cities where natural ice was not a viable commodity, especially medium-sized cities across the South such as Louisville, Birmingham, and New Orleans. The two kinds of ice directly competed in other cities, but there was not yet a big enough supply to destroy the market for natural ice.

In 1880, when the Census Bureau first examined the ice industry, the only ice machines mentioned that were not in breweries were located in the inland cities of the South. Out of forty different kinds of ice machine in operation around the world at that time, less than six found any favor in the United States. One typical machine produced only fifty tons a day, an extraordinarily small amount compared to the size of the American ice harvest at that time, which was about eight million tons per year (when the weather was good).[39]

By 1900, artificial ice production had become efficient enough, and therefore the ice it produced had become cheap enough, for this product to break into northern markets where natural ice had always dominated.[40] That domination had been based on geography and the poor quality of the ice that ice machines produced. By 1900, a US Census report concluded that ice "made according to the improved methods in use is as hard and lasting as any that can be obtained from natural sources."[41] By 1915, there were around five thousand such plants in the United States, producing an annual total of approximately twenty-six million tons of ice. Each year saw the establishment of between 300 and 400 new facilities.[42]

You can see the effects of the growth of artificial ice in the actions of New York's Ice Trust after 1900, when it began to fight for its life both politically and economically. The trust began to expand into other cities after the turn of the century, and when it entered these new markets, it built ice-manufacturing facilities. In Washington, DC, for example, the trust owned five of the nine ice-manufacturing plants operating there in 1906.[43] As ice machines became smaller and more businesses started making ice, it became increasingly difficult for the Ice Trust to maintain power over the industry except in New York City, where the Hudson ice crop was large almost every year. With ice being both harvested and manufactured, supply increased and prices generally dropped over time.

After 1900, the popularity of ice machines skyrocketed largely because the many additional advantages of improved artificial ice making proved difficult to resist over time. For example, ice manufacturers felt confident enough to manufacture what they called "Raw Water Ice," the name for ice made from undistilled water straight from a city's municipal water system. The first such plants appeared in the United States in 1911, and within five years they became the norm.[44] Because these plants were cheaper to build and cheaper to operate, they played an important role in gradually eliminating the remaining market for natural ice. The elimination of distilling equipment in ice factories was indicative of an important difference between natural and manufactured ice from a production standpoint.

The gradual move from steam to electricity helped make smaller ice plants more efficient and therefore more profitable. The disappearance of the need for generating equipment left much more space for the machines, which were already shrinking in size. To set up a large-scale, southern-style ice plant in New York City in 1890 would have required three thousand square feet of floor space for the ice machine, plus space for the steam engine, a boiler, and two wooden tanks to hold molds for the ice blocks that the machine produced.[45] Once smaller plants became viable, manufacturers didn't have to pay for so much space.

Electric motors were themselves getting smaller as the electrical grid improved.[46] This further increased their appeal. As a result of this change, the cost of making ice by artificial means dropped considerably. Previously, that cost could make up as much as 50 percent of the cost of making ice.[47] After 1900, electric utilities had more power than they could sell, and the price of ice dropped as a result. Similarly, the advent of raw water ice allowed ice manufacturers to get rid of all the equipment that had once been necessary to purify water from steam. This included boilers, heaters, distilling apparatuses, and feed pumps.[48]

After 1900, electrical generating stations started producing ice as a side business using the surplus power that they could not yet sell. By 1910, there were 122 companies nationwide making what became known as "by-product ice."[49] These ice plants received special rates to make their ice during off-peak hours.[50] By the 1920s, electrical utilities were giving customers special rates for purchasing refrigerators in the expectation that these power-hungry appliances would help soak up their considerable excess supply of current.

As ice machines got smaller, less expensive, and more efficient, they began to replace ice delivery in a wide variety of commercial industries. Producing your own ice effectively eliminated the cold chain for ice at that facility. The kinds of facilities that used ice machines included meatpacking plants, butcher shops, cold storage facilities, and grocery stores. Places such as restaurants came up with new uses for these smaller machines. For example, Horn and Hardart, the famous chain of automat restaurants (put a coin in the slot and take out your plate), started buying their own ice machines around this time because they became convinced that they were losing too much money because of melting. Each store got its own unit, which could produce 250 tons of ice per day (although they didn't have to run it up to its full potential every day if they didn't need that much ice). Not only did they save money on refrigeration, but the ice from the machine also helped keep their perishable food colder than natural ice could have.[51]

At the same time, entirely new markets developed for refrigerating equipment where ice had never been used before. A 1910 story in the refrigerating trade press cited an expert count of 143 commercial uses for refrigerating machinery.[52] These establishments included steel-making plants, florists, and even morgues.

"At first it may seem that refrigeration could have little to do with baking bread," explained a Columbus, Ohio, baker in 1908, "but as a matter of fact it is the refrigerating machine that gives the baker the control over the processes of bread making requisite to the ensuring of the all important uniformity in product now demanded by the customer."[53] Ice could never create that kind of cold. Indeed, artificial refrigeration could create a far lower temperature than ice alone, which explains why all these new industrial uses opened up with the perfection of this technology.

While people had been predicting that artificial ice would completely replace natural ice since the 1890s, it was clearly happening by the mid-1910s. Artificial ice might have been more successful in New York City earlier, but the Ice Trust deliberately impeded the spread of ice machines because they had so much invested in keeping the Hudson ice harvest. They did not want their extensive natural ice business to become obsolete. The Ice Trust never made use of ice machines in New York. Instead, it bought up the largest ice fields, in order to maintain its monopoly, and then fiercely defended that position.[54] That strategy would only work for so long.

The market for natural ice started to shrink as early as 1904. The number of ice plants had increased from 787 to 1,320 in just the first four years of the new century.[55] By 1914, there were 2,543 ice-making establishments across the country.[56] By 1919, there were 2,867. The value of the ice manufactured at those establishments almost doubled over that span, probably because of demand created by World War I.[57] In 1907, America still consumed around fourteen or fifteen million tons of natural ice, almost three times as much as it had consumed in 1880. The natural product lost ground rapidly after 1910.[58]

You can see that change by looking at the geographic spread of ice plants too. In 1899, thirteen states reported to the Census Bureau that they had no ice plants located in them. (They were all in the North, as that was the region of the country where natural ice could be obtained most cheaply.) By 1904, that figure had dropped to nine. In 1909, that figure had dropped to seven. In 1914, the only states that had no ice plants were Maine, New Hampshire, Vermont, and North Dakota.[59]

Meanwhile, not only was the number of ice plants in the United States growing rapidly, but the amount of ice they produced was growing rapidly too. Between 1909 and 1914, the total amount of ice manufactured in the United States grew from a little over fourteen million tons to over twenty-one million tons.[60] By

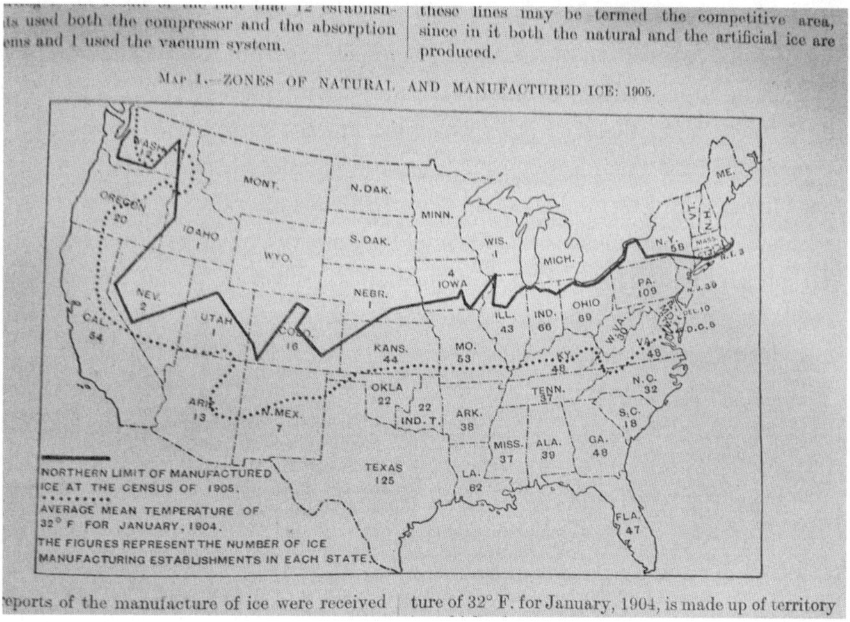

Map of zones where natural and manufactured ice dominated in 1905. As ice manufacturing became more efficient, the northern limit that manufactured ice could compete against natural ice moved farther north. From US Bureau of the Census, *Census of Manufactures: 1905*, 47.

1919, that figure had increased to almost twenty-eight million tons. If anything, argued the Census Bureau in its 1919 report on the industry, cost increases during the war slowed the momentum of growth in this industry rather than contributed to it.[61]

No wonder the natural ice industry entered full panic mode by the mid-1910s. A pamphlet from that period produced by the Natural Ice Association of America, entitled "The Handwriting on the Wall" and intended to raise funds for a pro–natural ice campaign, illustrates this panic well. "YOU ARE MENACED!" it begins. "So long as the people of your community are in ignorance of the special advantages of natural ice, just so long will capitalists continue to invest money in ice making plants. For every ton of output of such plants you lose at least Two Dollars."[62] Whatever special advantages natural ice possessed, they could not compete against clean, plentiful, cheap, and easily accessible ice made by machines.

The last great surge of natural ice collection in the United States came during the ice shortage caused by World War I. "The handwriting is on the wall," wrote the refrigerating engineer John E. Starr in 1920, echoing the language of the National Ice Association of America's pamphlet, "and not many years will pass when natural [ice] will be a memory, related in urban localities to the realm of the kerosene lamp, the hand-power machine, and the horse-propelled vehicle."[63] With natural ice becoming increasingly unprofitable despite many improvements made in harvesting technology since 1880, artificial rather than natural ice began to keep most people's iceboxes cold. While the practice of ice harvesting continued sporadically as a form of rural antiquarianism, few if any people lamented its inevitable passing. Perhaps that explains why the effective demise of the natural ice industry actually improved the day-to-day lives of individual consumers.

Transporting ice presented a different set of problems than finding a market for it. Ice, of course, melts. Therefore, its very existence was always a race against time. Ice could serve two purposes when it was on the move: it could be the object of the trip, or it could simply serve as the refrigeration for other things that were headed for market—usually perishable food. This ice traveled farther than ice destined for human consumption, and it persisted as a tool in the cold chain much longer because it took many years before mechanical refrigeration could shrink down so that it could efficiently keep a railroad car or a truck cold. For that reason, companies tied to food production needed access to ice far longer than American consumers did.

3 How Ice (and the Perishable Food It Preserved) Made It to Consumers

IN THE EARLY YEARS OF THE ICE INDUSTRY, the largest markets for the product were located far away from the cold regions of the country where ice originated. Because Frederic Tudor and his competitors harvested the frozen water of New England ponds and streams, cut it into large blocks, and transported it to distant places like Bombay or Savannah where the summers were long and hot, the losses due to melting were astronomical. Shortening the cold chain—whether by shipping it somewhere closer or by manufacturing it closer to where it would be used—automatically increased the supply of ice because it had less time to disappear.[1]

As ice became more popular in most of the United States, supplying these markets over shorter, more efficient cold chains became much more lucrative than selling ice overseas. Whether through increased production or less melting, more ice generally meant lower prices (unless there was a company in a particular market artificially controlling supply). Much of this ice for domestic consumption went into people's drinks, just as the well-traveled ice had done in the decades preceding this change in the industry. Abundant, cheap ice of local origin meant that more of the products that depended on ice would reach consumers and could be kept longer once they were purchased.

With more ice available, people used ice for new purposes, most notably to keep perishable food fresh. Of course, perishable food could travel without ice; it simply couldn't go as far without much of it spoiling. Ice extended the life of perishable foods of all kinds. The more ice that farmers had available to them, the cheaper it was to keep perishable food fresh and the fresher that food could be. The more ice that shippers had available to them, the further that food could be transported. The more ice that consumers could buy for their iceboxes, the longer they could keep that perishable food, and the more they could buy. This was true of ice that traveled by cart, by train, or (eventually) by truck.

Ice has always been very hard to move. Bulky, heavy, slippery—it was actually harder to move in the late nineteenth century than it is today because back then the cold chain had gaps in it where ice would inevitably melt. When it was not the commodity on the market itself, all the space that ice took up meant less space for whatever else someone was interested in shipping, such as milk, beer, or produce. Meatpackers first created their own cold chains in order to keep their perishable product fresh, but innovations that were developed for that product soon spread to the supply systems for perishable foods of all kinds. While ice shipped overseas earlier in the century provided its own insulation, ice that was supposed to keep something else cold required constant replenishment.

Before the perfection of mechanical refrigeration, it took a large supply of ice to make the entire cold chain possible. Gradually, as mechanical refrigeration improved, different parts of the cold chain began to use mechanical refrigeration. This improved the efficiency of the cold chain and lowered its costs while making accessible to consumers around the world all kinds of foods, even if they were out of season or could not be produced locally. In America, perishable food could now travel from where it was abundant to places where it was scarce. The mass production of agricultural products also became possible when the cold chain became efficient enough that large surpluses could be marketed over a wider geographic area.

The ice for New York City started its journey to consumers traveling by water. The vast majority of that ice came from the Hudson River, north of the city. There were about two hundred icehouses along the banks of the river south of Albany as late as 1910, employing twenty thousand men each year during the harvest season from January to early March.[2] When they finished cutting an ice field on the river, those men hoisted the ice they cut onto canal boats that looked like big, long boxes and sent them down the Hudson together in packs.[3] The Knickerbocker Ice Company, New York's largest and a forerunner of the Ice Trust, eventually owned sixty barges that could each carry nine hundred tons of ice, along with six steamers that could hold fifteen hundred tons of ice apiece.[4]

A fleet of ice boats arrived at the piers of the city every day during the summer. (It was harder to get ice to New York during the winter since the river froze up for much of that season.) Once the barges reached the piers, they docked next to a long platform known as an "ice bridge," a loading mechanism for the ice wagons that arrived at the piers each morning in order to get the ice they would sell that day. "The bluish white chunks are lifted from the boats by steam or horse power, and slid from the ice bridge into the wagon," explained the *New York Tribune* in 1901.[5] The wagons then delivered the ice directly to customers.

Ice delivery was a very complicated business, particularly in New York City, where there were few icehouses because of the high cost of land and the ample supply of ice just a hundred miles north.[6] In 1895, there were approximately fifteen hundred wagons and three thousand horses delivering ice in New York and the still-separate municipality of Brooklyn.[7] Ice wagons were like any other wagon, except they also had heavy, braced sides, a canvas-covered top with the name of the ice company inevitably written on the sides, and a tailgate to help icemen get on and off the back.[8] "The same very much soiled and luridly painted wagons that prompt thought of a defunct circus that were used twenty years ago for the most part continue to offend the perspective of nearly every city in the country," complained the *Washington Evening Star* in 1906. Some of these wagons could hold a few hundred pounds. Others were big enough to carry several tons of ice.[9]

Carrying so much ice in very hot weather led to the common problem of the horses that pulled these wagons dropping dead by the score. Even those horses that survived a season pulling an ice wagon were damaged by the experience. Unable to do this task two years in a row, horses sold at auction in the fall often went for a quarter of the value that they had been worth in the spring.[10] By the 1910s, ice trucks had entered the business. Some of these trucks simply went out to supply wagons already on their routes, while others actually delivered ice. They could, of course, carry much more ice than a single wagon, and they made it economical for companies in smaller cities to deliver all the way out to the suburbs or rural areas for the first time.[11]

The large, inevitably burly gentlemen who delivered ice to customers were known as "icemen." Icemen had an iconic place in American culture for many years—"the theme of song and story"—because of their daily feats of strength all summer, every summer.[12] They were symbols of male virility, particularly since they often visited housewives while their husbands were out working. While some women criticized them for tracking dirt on their living room floors, these

Loading deck at the Keystone Cold Storage Company, Reading, Pennsylvania, 1900. Nickerson and Collins Refrigeration Photographs, Archives Center, National Museum of American History, Smithsonian Institution

men nonetheless earned respect from anyone who saw them do their difficult work. This included loading their wagons each morning as quickly as possible so that their product would not melt and delivering very large blocks of ice to customers who lived up several flights of stairs.

Ice deliverymen generally worked for ice companies or ice delivery companies that bought their ice from the firms that either harvested or manufactured it. During the natural ice era, large conglomerates like New York's Ice Trust were vertically integrated companies that controlled everything from the harvesting process to bringing ice to people's doors. During the artificial ice era, the cold chains got considerably shorter. It was generally believed to be uneconomical to haul ice more than a mile from an icehouse or an artificial ice factory. Therefore, most cities contained a number of small manufacturing facilities or multiple points along the docks where barges of ice could be unloaded and shipped straight to consumers.[13] Service from ice companies in rural areas was always spotty unless you happened to live right next to a plant. Cities as large as New York also had some independent ice delivery companies that would buy excess ice from plants or brewers and peddle it to people who did not have contracts with the larger firms, like tenement dwellers willing to make special purchases on very hot days.

Commercial customers got a discount for buying the worst ice. Taverns, for example, used their ice to keep bottles and barrels cold. Since the ice had no contact with the actual drink, it didn't matter to them how clean that ice happened to be. Commercial customers also got a discount for buying ice in bulk. After servicing local businesses that depended on ice, icemen with regular delivery routes checked with their residential customers to see how many of them needed ice. Those customers who wanted a delivery would leave a card in the window.

Some cards had the different amounts of ice they wanted written on the edges, and the customer would leave the side with the right amount facing up. Other cards simply spelled out the terms of the contract for that year. For example, the Carpenter Ice Company of Philadelphia offered families and offices eight pounds daily for 45 cents per week in 1882. For 10 cents a week more, customers could get four more pounds of ice each day. They could also get twenty pounds for a single delivery.[14] Leaving the card in the window signaled that irregular customers wanted ice that day. When the wagons ran out of ice during the day, they would return to the pier (or, later, the ice plant) and get more. On a hot summer day, the iceman might not be done until ten o'clock at night. On colder days, they might be done by three o'clock in the afternoon.[15]

Large companies assigned icemen regular routes through their home cities. Early on during ice season, they would canvass every house along that route for regular customers willing to sign contracts for daily or weekly delivery. On each visit during the regular season, the icemen would cut however much ice the customer wanted off the larger block in the back of his wagon. (Cutting ice out on the pavement made the ice dirty; cutting it in the customer's kitchen left a huge mess of ice chips for them to clean up.) After weighing it, he then carried it inside on his shoulders (which would be protected from the cold by a blanket) or with a huge set of tongs. At some point between the 1890s and the 1920s, it became common for larger companies to employ a second, lower-paid person designated a "helper." This allowed the cart to service both sides of any given street at the same time.

Paul Angle worked for an ice company in Mansfield, Ohio, for four years during the late 1910s and early 1920s. He later remembered, "We left the company's

Iceman delivering Kennebec ice to schoolchildren in Washington, DC, ca. 1899. Library of Congress

barn at 6:00 a.m. on a hot Saturday and reached home at 9:00 that night, tired, hungry, soaking wet . . . possessed of the knack of carrying 100 pounds on the left shoulder and lugging 50, suitcase fashion, at the right side." Angle's cart held between twelve and sixteen three-hundred-pound cakes at a time. After taking care of the commercial customers and the wealthier families with ice contracts, his cart would take a leisurely journey through working-class neighborhoods so that he and his companion could sell to people without contracts.[16]

On Tuesdays and Fridays, Angle made an extra run covering just butcher shops. Each shop had a wooden icebox that could hold a ton of ice in its overhead compartment. "Had the ice chambers been empty butcher shops would have been a cinch," Angle later wrote, "but the bottoms of the ice compartments were always covered with a generous coating of slime and with the melted remnants of the last loading. . . . All these melted fragments had to be stacked at the back of the compartment." One of the icemen had to crawl into the compartment to do this, despite the risk of the remaining ice falling on him.[17] This kind of customer service suggests the extraordinary competition in the commercial ice trade.

With many different companies serving the same areas and ice generally the same price in every locale, ice companies had to compete largely on the basis of best satisfying the needs of their customers. That explains why Philadelphia's Washington Ice Company wrote on its 1883 card, "None but obliging Drivers employed."[18] This ideal seldom happened. As the only published treatise on the subject of ice delivery concluded, "It is utterly impossible to please everybody, therefore it is necessary that the operations of the delivery department will be such as will bring forth the least number of complaints and produce the greatest number of satisfied customers."[19]

Over time, icemen got a bad reputation for cheating both their customers and their employers. It was not hard at all to shortchange a customer. For example, you could claim that a nineteen-pound block actually weighed twenty; after all, it was not as if anybody kept a scale that big around. "The icemen bring the scales," explained the *New York World* in 1889, "and under protest will weigh the morsel of ice upon delivery, but the plain evidence of the senses is that in many instances their scales are base deceivers." Customers who complained about such treatment might as well have been talking to stone walls, the paper explained, since the icemen had an unspoken pact of not stealing each other's customers.[20] No wonder the iceman became in the public mind the embodiment of a large corporation (oftentimes a monopoly) that was charging too much for an essential consumer good.

Stories of such treatment obviously did not mean that all icemen were crooked. Delivering ice was an inexact business. Indeed, it was impossible to cut large blocks of ice to exact pound measures with the iceman's regular tools, such as an axe or an ice pick. However, you could make good money shortchanging customers or by selling some extra ice to customers on the side, chalking it up to melting and pocketing the proceeds. But icemen took lots of abuse for situations well beyond their control. After all, they were the only representatives of the ice companies that most customers ever saw. Therefore, they felt the brunt of complaints about price or supply or the fact that they arrived later than usual on days that were particularly busy. In 1906, one Washington, DC, iceman complained to a reporter that out of forty-two houses that he had visited one day the previous week, he had been abused at thirty-one of them.[21]

Smaller ice peddlers not associated with larger companies were a particular problem for the reputation of ice companies everywhere. Since the peddler had no contract or long-term relationship with his buyers, he had no incentive to treat them honestly. He could also blame the price of ice on the ice plant whether it was the fault of the ice plant or not.[22] Despite these competing interests, ice companies throughout the United States sold to the peddlers whenever they had excess ice because that was the easiest way to get a return on the money that they had already sunk into producing their ice. After all, they wouldn't get anything back if they just let the extra ice melt. Peddlers who worked independently often had worse reputations than icemen did, allegedly storing their ice in dirty cellars under ratty blankets while they waited for customers.

The way for larger companies to counter both kinds of cheating was for ice firms to get their drivers to sell their customers coupon books in advance of delivering them ice. In other words, pay for the ice in advance, get a small discount for doing so, and hand the iceman a coupon instead of cash when a delivery arrived. At the same time, the driver kept a book of his own in order to record purchases and deliveries. This not only limited cheating but also vastly simplified the ice industry's accounting and gave companies a much better handle on the daily activities of their employees.[23] These coupon books, as one observer wrote in 1918, "reduce labor and bookkeeping, minimize losses, control credit limitations, and give you an absolute check on your deliveries."[24] How the icemen felt about the changes brought on by coupon books is not readily apparent from the available literature. It is not to hard, though, to imagine an honest iceman being upset about this kind of oversight.

Sometimes the iceman himself would get cheated. "In several instances," reported the *New York Tribune* in 1891, "it has happened that an ice cart going over an uneven road noisily has been 'raided' by a gang of boys, and the entire load of ice fell into the street, one cake after another, without the driver being aware of it until too late." The boys would then take the ice they could recover home. "So bold had these predators become that it became necessary to station an extra man at the back of the cart when going through some of these neighborhoods."[25] Raiding an ice cart was actually an iconic scene from turn-of-the-century urban life. In 1906, when the city of New York was paying for free ice for the poor, an ice riot erupted. While the one policeman stationed there went to call for backup, the people in line cleared out the whole depot and disappeared.[26]

Misguided customer perceptions about the ice business were another reason that ice deliverymen had such bad reputations. For example, ice deliveries in cities where people were concentrated in one place could arrive faster and were more cost-effective than in towns where icemen had to deliver on routes that contained widely scattered suburbs. Customers did not factor in geography when considering the price of ice they bought, and therefore they accused the ice companies of robbing them blind even if the deliveryman weighed every block of ice perfectly. Similarly, people complained that the ice wagons did not deliver on Sunday mornings (when at least some drivers went to church), because they did not want to order more ice on Saturday afternoons (for convenience sake). Customers asking for special orders, such as when they were entertaining, raised the prospect of throwing off a company's carefully timed delivery routes.[27]

Perhaps the most important aspect of maintaining good service was for icemen to stick to regular schedules. Unlike the milkman (who left his product on the doorstep), the iceman almost always had to enter the houses on their route; hence, his visits had to be scheduled, so that housewives could know when they could be out of the house. If an iceman ran out of ice before a scheduled delivery or, even worse, forgot a regular customer, that customer risked having all the food in their icebox spoil. Customers who forgot to leave their ice card out when they actually wanted ice risked the same fate. Therefore, the most accommodating ice companies set up special emergency runs for the people with whom they had long-running relationships. While housewives with special demands drove ice deliverymen crazy, commercial customers were always much easier to accommodate.

Ice traveling in refrigerator cars full of meat was the first important use for ice in America beyond dropping it into drinks. Meatpacking companies needed it both for keeping slaughterhouses cool and for shipping dressed (meaning pre-slaughtered) meat over long distances. Before they used ice, meatpacking companies sent live cattle from Chicago directly to the East Coast. Not only was this extraordinarily inefficient, since it meant paying to ship parts of the cow that nobody would eat, but it also made for pretty bad-tasting beef. "The poor beasts are herded close together, cramped, bruised and lean," explained an article in *Scribner's*. "In a journey of over a thousand miles, occupying five or six days, they have been fed and watered only twice."[28] The underweight animals would then be slaughtered at abattoirs in the cities where they would be consumed. The decision to centralize the slaughter of animals in the Chicago Stockyards allowed meatpackers to use ice refrigeration to chill dressed meat on its way east. This proved much more efficient than the previous system for reasons of scale, let alone all the waste that ice prevented.

There were numerous attempts to devise a successful refrigerated railcar for dressed beef until Gustavus Swift (with the help of engineer Andrew Chase) perfected this practice during the early 1880s. The main design problem when carrying meat by rail was to separate the product from the ice. If the ice touched the meat, the meat would turn black. Over time, cars carrying meat began to employ brine tanks at each end. The tanks had separate compartments, with the ones full of brine closest to the meat. These kinds of cars could carry five thousand pounds of ice, as well as the brine. Another problem that such cars faced was to protect them from damage by all the water that came from melting ice. For this reason, steel gradually replaced lumber in almost every part of these cars.[29] The original cars that Swift had designed required the doors to be opened in order to replace the ice. This raised the temperature in the car, which hurt the meat, so Swift's final design allowed crews to reload the cars from the top.[30]

However, Swift did not stop with improvements to railcars. In order to get his product all the way from Chicago to customers on the East Coast, he had to buy storage space along the way and even set up routes for smaller "peddler cars" that served customers outside of big cities. Swift also had to hire sales and delivery staff to get his meat to butchers, grocers, and other retailers. Overseeing this large network required new administrative structures that weren't present in other companies of the era.[31] Butchers who once did the slaughtering themselves now cut up 200- to 250-pound "quarters" of beef based on local tastes and displayed the results in coolers that were also chilled by ice.[32]

As complicated as maintaining this network could be, acquiring ice to keep the meat fresh added another dimension to the supply chain problem. "He [Swift] had to buy ice-harvesting rights in lakes all over northern Illinois and Southern Wisconsin so that he might have the ice for cooling his beef and loading the ice boxes of his cars at Chicago," Swift's son later explained. "He had to develop icing stations all the way across the country to his markets in the East—the railroads would not build them. Then he had to get the ice-harvesting facilities to supply these stations. He had to build ice houses of huge capacity. His ice-consuming capacity was by the wave of a hand and the development of an idea greater than any other ice user's in the country."[33] As trains full of dressed beef traveled eastward, they would wire the stations to have half a ton of ice ready for each car when they arrived.[34]

All this infrastructure explains why meat was generally the most expensive food that anybody could buy. Swift had to buy 450,000 tons of ice per year during the 1890s.[35] Inspired by Swift's success, his competitors in the new dressed beef trade bought still more. This drove the ice trade farther and farther away from Chicago—not because the ice was polluted there (it was, but the packinghouses could have used it since the ice never touched the meat), but because the packers had driven up demand for ice so much that they had to find new sources for it.[36]

Slaughtering cattle had once been a seasonal activity. Packinghouses sat idle during summer months because the beef they could have slaughtered would have spoiled before reaching the market. Access to ice in the slaughterhouses made it possible for the Chicago packers to extend the slaughtering season. Access to iced railway cars made it possible for Chicago packers to supply meat to the masses year-round. The consumers of Swift's beef did not buy this ice, but still they benefitted from it. Dressed beef transported from Chicago to the East Coast in iced railway cars could not compete against today's mechanically refrigerated article, but it was a huge improvement to the quality of the cuts of meat which most people could eat before Swift started his experiments.

You have to understand what beef eating was like before ice came along in order to appreciate what ice did for the American diet. Beef has always been a food with a very strict hierarchy. Even today steak is better than round, round is better than chuck, and so on. The worst kinds of meat that people ate in the nineteenth century were much worse than the worse meat available today. Yet scraps cured in barrels during the mid-nineteenth century were still better to most Americans than eating no meat at all.

In this way, the existence of chilled beef changed every American's perception of how meat was supposed to taste. Pickled or canned meat tasted different because of the process used to preserve it. Despite ice's failings as a source of cold, dressed, refrigerated meat tasted much closer to the way meat tastes now. The results did not necessarily produce the best-quality beef, but it was better than beef that came from underweight animals or that had been preserved in other ways, like pickling or canning. Because the use of ice contributed to a great increase in the supply of meats of all kinds, it also contributed to an across-the-board drop in the price of meat over the course of the late nineteenth century.[37]

The use of ice along the cold chain changed America's perception of how fish was supposed to taste too. A fish begins to decay immediately after it dies because of the bacteria that were already in it when it was alive. On boats, the earliest solution to this problem was to ship fish back alive, but those fish ran the risk of hurting themselves during the trip back to shore. On shore, the earliest solutions to the fish preservation problem were salting or drying. However, both these methods drastically change the taste and texture of the fish. Canning dated from the early nineteenth century, but it took years to work out reliable methods that practically eliminated the risk of botulism. Equally important, canning also did not preserve the freshness of the product quite as well as later methods did.

That explains why late nineteenth-century fishing concerns began to experiment with ice. "Large quantities of cod, halibut and other salt water fish were once brought to market alive in the 'wells' of specifically constructed smacks," wrote the *New York Tribune* in 1900. "This custom has been almost entirely abandoned, the use of ice having proved more convenient to the average fishermen."[38] Despite the space that ice took up on fishing boats, the advent of mechanical refrigeration did not improve the quality of the fish preserved until flash freezing came along late in the twentieth century. The large machines available at that time were good for making ice cheaply, but never small enough to fit on boats that size.

Before ice became a commodity, only people who lived by seaports could eat fish regularly. Fish became a national commodity starting in 1857 as fish concerns began experimenting with shipping them inland packed in ice. One of the first breakthroughs in this trade was to learn to precool fish before shipping them out. This meant that the ice in the package only had to contend with the heat in the environment; it no longer had to contend with the heat of the fish. This saved on ice. Like meat, fish often traveled in refrigerated railway cars by the end of the century. Shippers generally used flour barrels for these purposes, with the fish kept between layers of coarse, crushed ice.[39]

As the price of ice dropped over the course of the nineteenth century, more fishermen were willing to take a chance on using it to expand the range they could take their boats and to improve the quality of their fish. Boats working the cod and haddock fisheries toward the end of the century generally carried between ten and twenty tons of ice on long journeys, but much less during the colder months of the year. Ice blocks went at the bottom of the storage pens on these boats. The fish went on top of that, followed by finely cut ice on top of the fish. Then came successive layers of ice and fish until the boat reached capacity.[40] In the twentieth century, fishermen operating in the Gulf of Mexico got their ice along the docks from machines that split blocks into chips. The resulting product emerged from a hose that could be used to spray the bellies of ships with as much ice as they needed.[41]

Ice proved absolutely crucial for preserving the taste of the fish being shipped from the boat all the way to the fish market and beyond. The faster fish were placed in ice, the longer they stayed fresh, and the better condition they'd be in when they reached market.[42] If you didn't carry ice on your boat to preserve your catch, you had to get home as soon as possible in order to get your fish chilled. This limited the range that any fisherman could cover because the farther you went out into the ocean to catch fish, the more your fish would decay on the way home. If you did decide to bring ice, you would either have less space for fish or need to acquire a bigger boat (with all its accompanying higher operating costs). Of course, this decision depended on the price of ice. Cheaper ice meant more ice could be used. More ice meant better refrigeration, which meant that more people further inland could enjoy fresh ocean fish.

Ice remained the main way to refrigerate fish during the late nineteenth and early twentieth centuries for the same reason that it is still used today: it is the best method available to preserve fish apart from freezing and the only commercially viable way to display fish for customers.[43] Icing fish not only cools it down quickly but also keeps the skin moist so it won't shrivel up (as can happen when it gets frozen badly). Unlike freezing equipment, ice has the advantage of being easily portable. While ice needs to be replaced, it can nonetheless keep fish fresh for weeks. Fishermen generally preferred natural ice to artificial ice, thinking that it was less cold and therefore more likely to melt, cooling the fish faster.

Fish is perhaps the most perishable food that Americans eat regularly. Produce, on the other hand, benefited from refrigeration but did not require it. Before 1900, railroads that shipped California fruit green faced an incredibly high spoilage rate because they rarely used ice. The business occurred only on a very small scale and suffered from numerous service complaints. Nobody wanted to pay for the infrastructure that ice required if they did not absolutely need it.[44] Northern California fruit moved between October and January and did not require ice. The Southern California crop required ice when it moved out during the warmer months in the rest of the country.[45] In 1894, the average iced railway car took five tons of ice initially and required fifteen tons total to make the entire journey to the East Coast.[46]

Before artificial ice came along, shippers had to obtain the rights to the ice along their transportation routes. These geographically induced monopolies had the potential to greatly increase the cost of shipping perishable foods over long distances. For example, because the Union Ice Company controlled all the icehouses around Truckee, it became the sole provider for refrigerator cars traveling along the original transcontinental railroad route across the Sierra Nevada

Mountains.[47] The advent of artificial ice broke that stranglehold by making ice available in places where it did not appear naturally. Lower transportation costs drove down the price of the perishable food, which in turn increased the consumption of California produce.

It also led to more investment in the infrastructure of California agriculture, so that it could grow further. The Pacific Fruit Express (PFE), a joint venture between the Union and Southern Pacific Railroads, formed in 1906. While they once bought their ice from private companies all along their routes, the company quickly built or purchased ice plants in order to vertically integrate their operations better.[48] The company handled forty thousand fruit cars per year by 1908.[49] Unlike the situation with meat, railroads maintained enough icing stations so that the fruit could be re-iced every twenty-four or thirty hours. While this kind of journey inevitably affected the flavor of the fruit, all of it could show up anywhere in the country without spoilage or other damage.[50]

Because the costs of supplying refrigerator cars dropped as ice machines got better and became more prevalent, the total number of cars the PFE sent quickly increased. By 1921, the PFE handled 142,000 cars of fruit per year. By 1923, the company handled nearly 200,000 loads. In 1921, the company manufactured 1.4 million tons of ice. By then, it had become the largest producer of ice in the world.[51] Much of that ice came at the beginning of the journey. In 1922, there was only one small mechanical ice plant in the Salinas Valley. By 1930, thanks in large part to better electrical infrastructure, there were six. This made it possible for much of the country's lettuce to be grown there.[52]

Ice and the railway cars it chilled had a similar effect all over the country. Consider the example of the Rocky Ford cantaloupe. Rocky Ford is a town on the plains of southeast Colorado which started growing cantaloupes during the 1890s using seeds imported from Massachusetts. Those seeds grown in that particular climate produced melons that were particularly large and flavorful. By 1900, those melons were shipped by the train carload to markets all across the Eastern Seaboard.[53]

The Rocky Ford melons were larger and sweeter than those previously available, and because they had a smaller seed cavity, they were far more efficient to ship. Moreover, because of the longer growing season in Colorado, cantaloupes became available for a month longer in New York City than they had been previously. The introduction of Rocky Ford cantaloupes in ever-increasing supply led to a huge drop in prices throughout the region. In 1904, at one point in the season, twelve to fifteen carloads of melons from Rocky Ford alone arrived in the city each day. The supply of melons in the city had increased by 100 percent over the previous year, and suppliers had to sell some of their product for less than the freight charges.[54]

A similar story can be told for other fruits besides melons. "Twenty years ago," explained a reporter for the *New York Sun* in 1894, even before the artificial ice trade really took off, "pears which sell to-day two for five cents were 40 cents apiece, and black Hamburg grapes . . . were only to be had from hothouse growers and were worth $1 a pound. The luscious great cherries which were so plentiful a few weeks ago were unknown in our markets, prunes were known only as a dried fruit, and apricots and nectarines were rare enough to be a tradition among the greater part of the people."[55] Variety would not only have improved the flavor experience of New Yorkers; the increased supply of fruits and vegetables would have lowered prices, since as sources of vitamins and nutrients they are, in a sense, interchangeable. More of any perishable food, therefore, made it possible for more people to consume it.

The effects of improved access to better produce brought by ice refrigeration on American health were immeasurable. The sociologists Robert and Helen Lynd did their best to describe them in *Middletown*, their famous 1929 study of Muncie, Indiana. "In 1890, the city had two distinct diets," the Lynds wrote: winter and summer. The winter diet had few fresh fruits or green vegetables. "Following 'winter diet' came 'spring sickness,'" they explained. "Nearly everybody used to be sick because of lack of green stuff to eat," explained a historian whom the Lynds quoted in their study. "What people ate in the spring when the greens returned had to cure them of the effects of their poor diet during the winter."[56] The seasonality of produce, in other words, was a detriment to good health.

Once ice made it possible to preserve crops from climes where they could be grown year-round, the people of Muncie no longer had to eat their traditional winter diet. "Today, more and more Middletown housewives are buying fresh vegetables and fruits throughout the winter." Besides the direct health benefits of this new diet to the year-round health of the population of Muncie, the variety of foods they ate would have made eating more pleasurable. It also saved women an enormous amount of labor in the kitchen because they no longer had to work hard to make the same limited diet seem more varied than it was.[57]

Because the American ice industry developed long before the advent of reliable mechanical refrigeration, consumers never saw much of the ice it produced.

That meant that the people who bought products preserved by ice could not always preserve those products well after they bought them. Despite this difficulty, having greater access to ice—even if it was just in the cold chain—changed what people ate. Ice changed other foods too, not just by transporting them more easily, but by changing the way they were produced and stored. And sometimes ice made life better without affecting the food supply at all.

4 How Ice Changed the American Diet and American Life

BY 1900, THE ICE INDUSTRY HAD BECOME thoroughly modernized. Output was high. Machines both produced ice and helped harvest still more. Ice machines were efficient enough that they had begun to spread to even the most northern parts of the country. These machines could supply ice on demand, whenever necessary. In short, the industry had practically overcome the limitations once imposed on it by the weather. Ice famines, which had periodically caused prices to skyrocket earlier in the history of the industry, became a thing of the past. With a surplus of ice, the product could be used in ways that people never would have imagined when it was relatively rare.

For example, without greater access to ice (harvested or manufactured), brewers never would have been able to keep up with the demand for lager beer. In addition to railroad cars filled with ice for transportation, the meatpacking industry used ice to chill their factories. Besides expanding the range of distribution for perishable products of all kinds, refrigeration cut waste due to spoilage. This increased the supply of all perishable products, which in turn lowered their price and therefore made it easier for more people to consume it. In some cases, the increased availability of ice also made those products safer to consume since they were preserved better throughout their journey along the cold chain.

Milk is an excellent example of this last impact. It takes less than two days for fresh milk to spoil without refrigeration. Dairy farms were much closer to urban markets at the turn of the twentieth century than they are today, but they were also much smaller and more numerous. That meant that milk had to be collected from a large number of farms and stored before it could be distributed through the middlemen who retailed it. All this milk required refrigeration along the way, or it would spoil before reaching the consumer. Before refrigerated trucks, that refrigeration generally came from ice.[1] A 1915 ad for an El Paso milk producer bragged that they were "the only dairy that uses ice machines to rapidly cool and preserve the milk." Those machines also stocked the dairy's delivery trucks with ice, keeping the cold chain as complete as possible until their product reached customers' kitchens.[2] Ice, and later mechanical refrigeration, filled in the holes along the cold chain for milk, keeping it fresh and healthier.

As these holes disappeared later in the twentieth century, not only milk consumption but the consumption of dairy products in general increased sharply. For example, between 1910 and 1940, butter consumption in the United States was four times greater than it has been in recent decades.[3] Without ice refrigeration, much of the milk needed to make butter never would have survived the trip to the factory. The more refrigeration, the safer that milk would have been. In this way, the increased availability of ice made milk both tastier and healthier than it had been previously. In fact, as with produce, the mass production of milk and dairy products was only possible because enough ice existed to preserve it on its way to distant markets.

Refrigeration and pasteurization both played distinct roles in the making of the modern milk industry, but both processes worked toward the same overall goal of increasing consumption. The effects of ice on milk distribution predated pasteurization, which gradually became the law on a municipal level starting with Chicago in 1908.[4] Bacteria in milk, particularly tuberculosis, thrived in heat, so refrigeration certainly made milk safer. However, since pasteurized milk could still spoil, refrigeration remained necessary in order to keep even healthy milk marketable. This explains why the pasteurization and refrigeration of milk became popular around the same time, between 1900 and 1920, as well why both practices remain essential to this day.

By making the cold chains for perishable products like milk more complete and effective, ice became the foundation for the complex modern food preservation systems of today—not only in terms of the technologies involved but also in terms of developing a market for a wide variety of foods that needed to be preserved. The ice provisioning system also made it possible for Americans to consume more ice itself in an interesting variety of ways. While what follows in this chapter is not a comprehensive summary of all the foodstuff that depended on ice during this period, it does cover many of the examples in which the ice industry took great pride. However, none of these foods could have developed a mass market if not for the development of one particular technology: what we now call an icebox.[5]

All concerns about the cleanliness and transparency of natural ice stem from the fact that people who drank iced drinks directly ingested the product. But ingesting ice was hardly the only use consumers had for that product. Ice always had to be stored, no matter its eventual use. After all, since ice melts, it would not last very long if left out at room temperature. This is not a serious concern in an era when just about every consumer in every developed country can keep ice in the steady, reliable cold of their home freezer. Before mechanical refrigeration made it into people's kitchens, iceboxes developed so that ice wouldn't melt quite as fast as it would have without them.

An icebox was any box designed to keep ice in it. Usually it had two compartments, one for the ice and one for the food. There were openings between those compartments so that cold air could circulate through the entire container, but the idea was to separate the ice from the food. Since ice—particularly natural ice—was often dirty and filled with sediment, separate compartments kept the food cleaner and prevented the ice from damaging anything being kept in the icebox. The first iceboxes appeared during the mid-nineteenth century as the ice industry became more popular. They were handcrafted appliances made by carpenters or homemade objects refashioned out of available wood, like that from barrels.

By the late nineteenth century, iceboxes became mass-produced, better designed, and, as a result, more effective. Higher-end models were made out of porcelain. Even lower-end models came with insulation, so the ice would keep longer. More air passages inside these newer iceboxes allowed for better circulation. As iceboxes became a fixture in American households over the course of the late nineteenth century, they began to take on the look of fine wood furniture. The existence of iceboxes not only created a new way for consumers to use ice but changed the American diet forever, since now it became possible for the first time to save leftovers.[6] This helped deal with the terrible problem of food waste. It also saved women cooking time and limited how often they had to go shopping. Providing easy access to ice in both urban and rural areas, no wonder iceboxes proved extremely popular in the United States.

On the other hand, iceboxes had many drawbacks compared to modern refrigerators. You had to remember to change the pan where the melted ice collected (or, later, clear the drainpipe), or otherwise you'd have to mop up the overflow. Iceboxes were difficult to clean, particularly since smells from the food inside could permeate the wood. If different kinds of food were kept too close together in an icebox, one would start to smell like the other. Many of these machines, even after the turn of the twentieth century, could not keep food any lower than fifty degrees Fahrenheit. They could do nothing to control humidity. If you opened them too often, the ice would melt. More importantly, if for some reason your iceman missed a delivery, all the food inside would spoil.

Of course, consumers also had to pay for all the ice these appliances used. Iceboxes kept food better when stocked with large blocks. "A large piece of ice usually lasts longer than an equal weight delivered at intervals in smaller pieces," explained a writer for *Good Housekeeping* in 1913. "Where the ice is allowed to run out, more is required to get things cooled off again than if a fair amount had been kept on hand."[7] The reason for this, as explained earlier, is that ice can only keep things cold by melting. Smaller ice chunks have more surface exposed to warmer air and therefore melt faster. Opening the door too often was another way to make the ice melt faster.

Because ice had to melt to work, trying to insulate ice so that it would last longer actually put people's food at risk. That explains why ice manufacturers hated one particular technology: the ice blanket. The makers of "Preservo Ice Blankets," for example, bragged that their product "keeps your ice from melting and increase[s] refrigeration." In truth, it "can be of benefit only when a very slight coolness of temperature is sufficient," explained the journal *Refrigerating World*, "and since most domestic iceboxes are not cold enough under ordinary circumstances, there is little use for it."[8] Ice companies understood this but did a poor job of explaining it to their customers; otherwise, such products would not have existed.

Iceboxes were also a good way for Americans to get what was a much-prized drink for them: iced water. Put water in bottles, store those bottles right next to the ice, and they would quickly become cold. This was the only thing that most home economics textbooks ever recommended storing in the ice compartment of an icebox, as the bottle protected the water from contamination by the melting ice. Iced water had many other uses besides being consumed directly. For example, some people kept it around as a way to weaken iced tea.

Those interested in really cold water could put ice directly in their water, or just wait for the ice to melt. This was not common during the natural ice era because of the sediment contained in it. However, drinking water that came directly from melted ice became increasingly common with the advent of manufactured ice since that ice was made from purer water. "If artificial ice is made with due sanitary precautions among the workmen employed," explained a doctor writing for *Ice and Refrigeration* in 1911, "there can be no possible objection to mingling it with drinking water."[9] The same could never be said for natural ice because nobody ever knew exactly what was in the water before it froze. As time passed and pollution became a bigger problem, the risks only increased.

This explains why the first thing that bars and taverns did after receiving blocks of ice in the morning was wash them. Even though much of that ice would keep beer barrels or bottles cold from the outside, the best establishments limited their customers' exposure to filth of any kind. That's one reason why the bartender's guide published by the world-famous Hoffman House recommended that "artificial ice is preferable when it can be obtained [since] it does not produce the same amount of slime and dirt as the natural product."[10] Starting in the 1880s, people began to fear that they could get diseases like typhoid fever from pollution frozen in ice. "An old friend indulged in a cocktail which contained ice frozen at Hunter's Point. As all intelligent people know, Hunter's Point is not entirely destitute of live pork and this live pork occasionally go swimming for its health," joked the humor magazine *Puck* in 1882.[11] Obtaining clean and clear ice made these entirely justified fears less likely.

People worried, of course, because much of the ice consumed at saloons, taverns, and bars went directly into the drinks. Cocktails were an early nineteenth-century American creation, and many of the early ones depended on ice for their novelty. As ice became increasingly common in American life, so did cocktails with ice in them. They first caught on in the United States during the 1850s, just as the natural ice industry expanded into more domestic urban markets for the first time. Indeed, one of the most important differences between the American and British varieties of punches, for example, was that the American ones had more ice in them, since ice was so much cheaper in the United States.[12]

By the mid-1880s, ice had become "the basis for all mixed drinks" in the United States.[13] There are four kinds that bartenders had to be able to produce from the blocks that got delivered every morning: block, lump or cubed, cracked, and snow. Block ice was ice just as it was delivered, but obviously cut into pieces that fit inside bar glasses. Bartenders produced cubed ice by shattering the block with an ice pick. Bartenders used an ice hammer to turn cubes into cracked ice.

Bartenders employed different kinds of ice shavers in order to create different kinds of snow (for a daiquiri, for instance). The larger the individual piece of ice in a drink, the slower it melts. This leaves the drink less diluted (but also warmer). Big square cubes from a refrigerator may last a very long time, but some drinks require ice that melts quickly.[14]

Bartenders produced these different kinds of ice using different kinds of tools. These included cocktail shakers, ice tongs (for handling the blocks), breakers, axes, mallets, and bags. (Place the ice in what is known today as a "Lewis bag," then pound it with a mallet, and voila! Crushed ice.) Most drinks of the time called for some kind of shaved ice in them, and the bartenders would do the shaving—often directly into the mixing glass. Bartenders also provided straws with their drinks because the dentists of that era thought that ice should have no direct contact with people's teeth.[15]

Ice surrounded bartenders in late nineteenth-century America, since nearly every drink called for it. Ice was also the first thing to go into a drink, after which ingredients like syrups or fruits would be added. Since people would watch the bartender exercise his art, this was another good reason for the ice to be clean. "The ice used for mixed drinks should be perfectly clear," declared one bartender's guide from the 1880s, "and when broken up it should be handled with an ice scoop and not with your fingers."[16] This may explain why bartenders from the nineteenth century appeared to be magicians to many observers, manipulating their ice for show, using tools that nobody else possessed.

How much ice patrons wanted depended on the region from where they came. A Brandy Smash, for instance, called for bartenders to "shovel" ice into the drink. It was popular in the South, but not around Philadelphia.[17] "Ice-ice-ice, you never get too much ice around a St. Louis bar," complained a well-traveled out-of-towner quoted in that city's *Globe Democrat* in 1882. "The St. Louis cocktail is an offspring of the river boats, and they get that, of course, from the Gulf City"—meaning New Orleans, where ice machines were common even at that time. "Ice-ice-ice. You can never get enough ice in a Mississippi River man's drink."[18] Without easy access to ice, this kind of predilection never could have developed. Cold cocktails simply would have been too expensive for the vast majority of Americans.

Besides consuming more ice, Americans conceived of more varieties of cocktails because they had so much ice with which to experiment. "Now, [the] mint julep, and sherry cobbler, and gin strog and other similar decoctions were not decoctions at all without ice," explained the journal *Ice and Cold Storage* in 1905.[19] As early as the 1860s, almost every cocktail in America used ice in some way.

Sometimes the drink was strained over ice. Sometimes the ice went directly into the drink. Sometimes bartenders used ice to chill the glasses that the drinks came in, rather than the drinks themselves. Sometimes the cocktail required iced water.[20] A bartender's guidebook from 1900 included many different kinds of ice for its colder cocktails: a "lump of ice" for a few of its highballs, a "long cube [of] ice" for its burgundy punch, and a "goblet of fine ice" for a Tom Collins.[21]

The mint julep is a good example of a popular drink that would not have existed in its modern form without the existence of an ice industry. The first American drink to become famous worldwide, it was a product of hot American summers and the easy availability of ice—especially in the South, where it originated. In its early days, a mint julep consisted of ice, brandy, water sugar, and "four or five stalks of unbruised mint." It's no coincidence that ice was the first ingredient here. Not only did ice go into the glass first, but by the late nineteenth century these drinks were served with ice crusted on the outside of the glass too.[22] Even though the mint julep predates the existence of the ice industry, easy access made ice an integral part of the drink.

Like the ice in cocktails, the ice in ice cream became part of the final product. Ice cream derived from the pre-nineteenth-century practice of flavoring ice. Rather than just adding a sweet syrup to crushed ice, ice cream includes cream, fruit, and other goodies, and the result is something that has proven incredibly popular for centuries.[23] Ice was both a component of ice cream and, like the role of ice in preserving milk, the way that ice cream makers kept ice cream cold. Before the ice industry came along, the only way to get ice cream cold enough to freeze was to use an ice and salt mixture. This creates a temperature below thirty-two degrees Fahrenheit, which could freeze all the ice cream ingredients together as long as they were constantly stirred. Before the nineteenth century, only confectioners made ice cream because of the high price of ice and all the labor needed to keep the freezer shaken.

Making ice cream depended on ice cream freezers. At first such mixtures were shaken. With the introduction of the hand-crank ice cream freezer in 1843, it became possible to make ice cream at home. Take turns cranking the freezer, and it could even become a party game. Whether shaken or hand-cranked, ice cream produced in either of these ways only came out in small batches. When ice production increased, particularly after mechanical refrigeration became common, it became far easier to mass-produce ice cream—particularly as the stirring process became increasingly mechanized. Some of the ice cream freezers developed during the late nineteenth century were powered by steam, allowing them to operate at a much larger capacity. These devices became the foundation of the American ice cream industry.[24]

Old-style ice cream freezers remained in use into the twentieth century for making ice cream at home. Ice cream factories depended on a number of new technologies. When ice cream was made at home, the ice that went into the mixture had to be hand-pounded. When a Minnesota reporter visited a new ice cream factory in 1890, he noticed "a dozen mammoth cakes of ice . . . which were evidently to be made into a frapped hash by a machine which with its iron teeth was industriously chewing into fragments a slab of ice that was held in its jaws." That was the ice that, when combined with salt, kept the ice cream cold during delivery. Another machine that mixed the ice cream itself produced "the music of crunching ice" for eight minutes as it combined all the ingredients.[25] The ice that went into the ice cream got crushed to a powder.

The advent of artificial refrigeration eliminated salt entirely from the ice cream production process, since one could then use mechanical means to make the ice and cream mixtures cold.[26] Nevertheless, early ice cream manufacturing still required ice as part of the mixture that would become the ice cream and for keeping the final product cold during shipment. The problem that ice cream plants faced, though, was that ice only came in very large blocks. Therefore, they employed ice-breaking machines, which could turn those blocks into small chips that would be used for either purpose. The manufacturing process at one New Jersey plant in 1894 required twenty-five pounds of ice in order to make about forty quarts of ice cream.[27] As mechanical refrigeration got better, bigger, more modern ice cream plants began to make their ice on the premises.[28]

When ice cream moved from a craft item to something that got manufactured, many people stopped making it themselves and started buying it at retail. But many people still made it themselves even at the end of the nineteenth century thanks to cheap ice. "Now there are perhaps twenty large firms and an almost endless number of small concerns engaged in this occupation in and around Manhattan Island," reported the *New York Sun* in 1891, "to say nothing of the thousands of households in which the ice cream freezer is an article of furniture."[29] Thanks to mechanization, those smaller firms might employ just one or two people to make ice cream. Those large firms could use as much as one hundred thousand pounds of ice per day.[30]

Without grocery stores, ice cream factories depended on street peddlers to distribute their products. These peddlers traveled through city streets during the late nineteenth century, distributing a product kept cold by cheap and abundant ice. After all, people didn't have freezers until household refrigerators came along during the 1920s. Therefore, the ice cream they bought had to be consumed quickly. Those drivers announced "their coming by the sonorous yells of 'I-c-e-Scream, I-c-e-Scream' which are apt to remind people awaiting the approach of the dinner hour that ice cream would not go bad for dessert."[31] In 1901, there were four thousand ice cream peddlers in New York City alone.[32] Many of the peddlers in New York and elsewhere were Italian immigrants, which was fitting, as Italy was the place where ice cream originated. Between 1910 and 1940, most Americans consumed their ice cream at drugstores and ice cream parlors because they didn't have freezers at home. Ice cream consumption peaked after World War II, when consumers could purchase it at supermarkets and keep it indefinitely in freezers once they brought their frozen treat home.[33] Before that point, cheap ice cream depended on cheap ice, which made it possible for ice cream to become accessible to the masses during this period. It has stayed that way ever since.

While ice continued to have an important role in the transportation of perishable food until at least the 1950s, creating practical mechanical refrigeration for cold storage took far less time. Indeed, any equipment that made ice could be easily modified to chill storage space instead. Nevertheless, ice was an important source for refrigeration in cold storage until approximately 1905, when mechanical refrigeration began to take over (even if the people who used mechanical refrigeration didn't figure out how to use it particularly well until after World War I).[34] Through lowered prices and increased availability, the food that cold storage kept fresh changed the American diet forever.

The first attempts to use ice for cold storage involved placing the ice in direct contact with food. That method worked only with foods that could stand up to water—like fish. Anything else would end up soaking and moldy before too long. Once people realized that the ice had to be separated from the food, they built two-story warehouses with the ice on top and the food on the bottom, with holes in between to let the cold air circulate.[35] In 1894, *Scientific American* described one such system in which the ice was "broken by hand power," a trap door got lifted, and the cold air wafted up toward the cold storage rooms through ten-inch metal pipes.[36] While these systems worked for a time, they did not remain effective, and none survived very long.

In 1901, the refrigerating engineer Madison Cooper began marketing a new ice-based cold storage system that used ice to chill brine, which in turn chilled the food. Mechanical refrigeration efforts at the same time did this to separate caustic ammonia from the food. Cooper did it to produce a dry cold that could keep food fresher. Cooper called it the "gravity brine system" because the ice was on top of the brine pipes, which remained in the air above the storeroom. In this manner, gravity kept the brine circulating, and the brine chilled the food.[37] In an article trying to sell his system to egg producers, he noted that "the ice for cooling has no more connection with the egg room than if stored in another state."[38] Accentuating that separation this way emphasized how his system prevented the kind of damage that ice could do if in direct contact with any food.

The most interesting thing about this system is not the system itself, but Cooper's claims for it. At the beginning of an important sales pamphlet, Cooper noted, "Ice has been placed in a position of an antiquated method of refrigerating, and this position is not at all warranted by present possibilities." More importantly, he noted that "successful cold storage on a small scale does not await the advent of a successful small ice machine, which is considered impracticable by our best engineers."[39] In another pamphlet from the same era, he wrote, "The requirements [for successful cold storage] have not been successfully met by an ice machine and probably never will."[40] Mechanical refrigeration progressed quickly after 1900, but when Cooper published those words in 1901, they still seemed like a safe prediction.

In 1904, Cooper published a textbook called *Practical Cold Storage*. While it purports to be an even-handed treatment of the subject, Cooper included many swipes at mechanical refrigeration as a source for cold in cold storage. "Even those who manufacture and install refrigerating machinery and apparatus often show either gross carelessness or ignorance of the requirements of a house which will produce successful results," he wrote in the book's introduction.[41] In fact, cold storage through mechanical refrigeration was viable by at least 1895.[42] Many of those first cold storage warehouses that used mechanical refrigeration were directly connected to and operated by ice-making firms.[43]

The same changes that made ice machines smaller, cheaper, and more efficient made cold storage by mechanical refrigeration increasingly viable. Cold storage facilities, such as ice-manufacturing plants, spread quickly as mechanical refrigeration equipment improved. Despite designing a cold storage system based on ice, Cooper wrote in *Practical Cold Storage* that "mechanical refrigeration . . . was much superior to ice refrigeration in that the temperature could be controlled more readily and held at any point desired and a drier atmosphere was

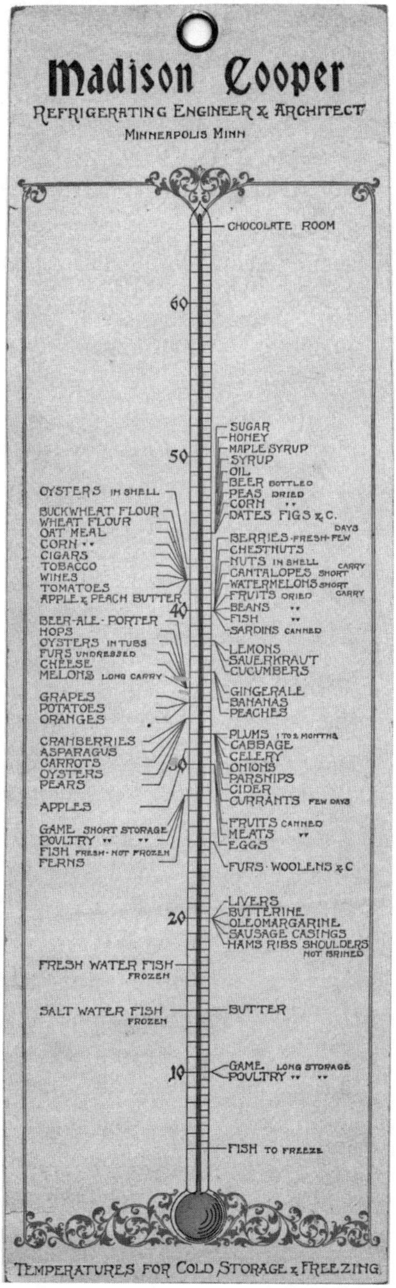

Cold storage thermometer illustrating the different optimal temperatures for different perishable foods. Mechanical refrigeration could offer the different range of temperatures needed to preserve different perishable products. Ice refrigeration couldn't. Madison Cooper Papers, Archives Center, National Museum of American History, Smithsonian Institution produced." He simply claimed that his system was best for small facilities.[44] In fact, ice could not take any space below freezing unless salt was applied to it, and this had the unfortunate effect of making it melt faster. Mechanical refrigeration, on the other hand, could take a space down to thirty-eight degrees below zero as early as 1897.[45] Mechanical refrigeration, in other words, was vastly superior to ice as the source in cold storage facilities in practically every instance.

This explains why warehousemen abandoned ice for use in cold storage as soon as possible. By 1914, only seven warehouses in the entire country still depended on ice for their refrigerating power.[46] Cooper himself would also make the transition to cold storage through artificial refrigeration. His career as a refrigerating engineer stretched all the way into the 1950s, but his initial resistance suggests that the replacement of ice did not seem inevitable, even if it proved simply to be a transitional stage on the road to the elaborate cold chains of today. Yet there are other ways to use ice besides consuming it or utilizing it to protect food.

Ice-skating is an activity (if not a sport) that goes back centuries. Obviously, early skating was only possible during winter when lakes or ponds froze. The formation of the Philadelphia Skating Club in 1849 led to a surge in the popularity of ice-skating in the United States.[47] New York City planners created the first artificial rink in the United States in the southwest corner of Central Park in 1858.[48] However, this was not an artificial rink in the modern sense of that term. Instead, the ice was flooded (as they did to improve the quality of natural ice harvesting), shaved flat, and continually repaired by park officials, since countless thousands of New Yorkers made use of this recreational opportunity.[49]

The advantage of an artificial skating rink was obvious to people inside and outside of the mechanical refrigeration trade—skaters would no longer have to depend on the weather in order to enjoy this activity. It is difficult to pinpoint the first such rink in the United States, although such a facility would have required substantial advances in mechanical refrigeration beyond what existed in the 1860s. The English inventor John Gamgee conceived of the first artificial ice rink in the modern sense of that word while attempting to improve the process of freezing meat so that it could be shipped in from Australia and New Zealand. The rink had a concrete surface. Atop that were layers of earth and timber planks, crisscrossed with copper pipes carrying refrigerant. These pipes were submerged in water, which froze when the machine started. The first rink in London opened in 1876, but it suffered from dampness and a thick mist as a result of the ice melting.[50]

By 1880, New York City had its first successful indoor artificial ice-skating rink. Built by a man named Thomas L. Rankin, it was located on Madison Avenue between 58th and 59th Streets. "We believe [Rankin] to be the first to succeed in artificially producing with economy large permanent sheets of ice for skating rinks," wrote *Scientific American* at the time of its opening. Rankin's system depended on chilled salt brine carried through iron pipes under a pool of water two hundred feet long and forty feet wide.[51] When not in season, Rankin's rinks could be used for other activities—such as bicycle racing, another fad from this era. Other rinks were funded by private clubs, usually for private skating or ice hockey games. The St. Nicholas rink on Lexington Avenue in New York City, a joint venture of the New York Athletic Club and the St. Nicholas club hockey team, cost $350,000 to build.[52]

By the 1890s, ice rink technology had been significantly improved. One rink described in the *New York Sun* in 1896 had small pipes full of chilled brine circulating to every part of the ice in order to keep the surface cold despite the heat generated by skaters and the chips made by the blades on their skates. "This brine," explained the paper, "is cooled by the same process used in cold storage warehouses and breweries and in the manufacture of ice." The surface of the rink was refinished three times each day, and each time it was "as smooth and even as if it had been perfectly polished."[53] Another rink described by the *Milwaukee Sentinel* in 1897 froze an entirely new surface twice each day, and "even the chips turned up by the skates do not melt when the hall is crowded."[54]

Despite such success, artificial ice-skating rinks still had limitations during this era. In another section of that same *New York Sun* article, the reporter interviewed the owner of the new rink. "Nobody feels like skating in August," he explained, "and the temperature necessary in order to maintain an ice surfaces so much colder than that outside that it would really be dangerous."[55] A good analogy here might be artificial snow on a modern ski slope. In other words, artificial rinks during the 1890s still required help from nature in order to exist. Trying to keep a smooth surface on an ice rink in warm weather was too tough on the equipment in order for the effort to be profitable. The Hercules Iron Works built an ice-skating rink along with the cold storage warehouse that it erected for the Columbian Exposition in Chicago in 1893. The two used the same equipment.[56] However, both facilities were lost when leaking ammonia refrigerant exploded before that fair even started.[57]

By 1910, thirty-five ice-skating rinks had been built around the world. Most of them were enclosed, since an open-air rink required substantially more refrigerating power and all of them were located in temperate zones where seasons changed, but all these rinks were only operational in the winter. The majority of these rinks operated on a brine system, meaning that the refrigeration chilled the brine, and the brine chilled the water, causing it to freeze. This drove up the cost of erecting a rink, which limited their spread beyond large cities.[58]

When mechanical refrigeration equipment improved enough to make brine systems unnecessary, so did the quality of the artificial ice rinks that that equipment could produce. By the early 1930s, using the direct expansion system pioneered in cold storage warehouses, the air temperature inside a rink could be between fifty-five and sixty degrees Fahrenheit without damaging the ice.[59] "Climatic conditions are no longer a factor in ice hockey, curling or ice carnivals," explained the journal *Refrigerating Engineering* in 1939, "and as a result northern sports have moved south. . . . Just a few years ago—ice sports were enjoyed only on natural frozen areas or on prepared rinks in the open."[60]

One improvement that made this situation possible were better piping systems with thinner pipes that could distribute refrigerant evenly across every inch of the surface of the rink. Another improvement involved the way that ice rinks were made. Instead of freezing a large pool of ice all at once, rinks were made by freezing ice in layers of water 1/32 of an inch deep at a time. While this sounds time-consuming, a modern ice rink during this era was only one and a quarter inches thick.[61] That suggests the precision and reliability that mechanical refrigeration had achieved by this time.

In hot summers, some people (especially children) bought (or maybe stole) ice just so that they could have something cold to melt on their foreheads. This sentiment, when extended to the commercial realm, explains the roots of air-conditioning in the United States. Chill water, and it becomes ice. Chill open air, and it becomes cold storage in a commercial setting and air-conditioning in public settings or at home. When ice was the only form of cold readily available, people used ice not only for cold storage but also for a primitive form of air-conditioning.

Before marketing his mechanical refrigeration machine to make ice, John Gorrie developed his invention so that he could cool his patients' rooms. An 1864 article in *Scientific American* proposed a hospital cooling system in which air would be run through cooling filters bathed in melting ice. The problem with such systems at that time was that ice was too expensive, which meant that they were not practical. Mechanical refrigeration was less efficient and even more expensive than ice, so that kind of cooling was not even considered until many years later. It was much cheaper to produce heat than cold.[62]

The most common form of early air-conditioning experiment was to blow air from an electric fan over ice to make a space cool. Even in the early days of mechanical refrigeration, this kind of arrangement had the advantage of not requiring a lot of capital since refrigerating machines were generally large and expensive. The disadvantage was that they were only temporary since ice, of course, melts. That fact also meant that there would be a puddle to clean up after the ice block disappeared. The vast majority of such experiments were undoubtedly never recorded.[63] For example, running an electric fan over a block of ice was a common household treatment for dealing with family members who were running a fever. The Navy built a particularly elaborate system along these lines for President James Garfield on his deathbed.[64]

The most likely place to find a written record of this kind of primitive cooling is when these arrangements were set up in theaters. Theater owners wanted to draw people in to see their shows even in summer, so explaining their cooling arrangements was good business. In 1896, the *New York Tribune* described the system at one theater during summer nights when "no known application of cooling agents is found too much." On those evenings, management placed ice blocks directly into the theater's ventilation ducts. "There is room for fifteen tons of ice," the paper reported, "but when it takes thirteen tons to keep the house cool for an evening, that may be counted as a reasonably hot night."[65] Multiply such numbers times the number of theaters in New York City, and you can imagine a lot of ice melting on behalf of theatergoers up and down Broadway. In 1910, the *New York Times* reported on a new theater that tempered its air by sending it through "ice chambers." As a result, "the most difficult problem of the mechanics of the theatre of cities of changing climate is solved."[66]

In the years before mechanical air-conditioning became common, New York theaters used ice machines to manufacture the ice that they melted to keep their patrons cool. "Down in the basement of the Palace [Theater], under the auditorium, is a giant ice machine that manufactures many tons of ice daily," reported the *Tribune* again, this time in 1914. "This ice keeps the big refrigeration chamber at an arctic temperature. Enormous blowers suck in air from the outside and suck it through the ice chamber. It is then distributed through the entire theater." The paper claimed that this was the largest such installation in any theater anywhere.[67] It was, at the very least, a much more complicated apparatus than what New York theaters were using for cooling just two decades before.

As was the case with so many other examples of ice cooling, mechanical refrigeration chilled things or kept things cool more easily and better. In 1890, the

New York Herald asked, "Is it anticipating too much that the New Yorker of the near future will be able to bring the temperature of Newport, Cape May or even of Greenland into his office or his parlor or even his bedroom by merely touching a button and fixing a governing key on the dial beside his desk or chair?"[68] In other words, even uninformed observers understood that the same technology that could freeze water could be used to chill people. Perhaps the most interesting aspect of that particular question is the way it poses the possibility of lowering the temperatures of people's homes rather than giant theaters. While nobody wanted to sweat through a night on Broadway, most New Yorkers probably did not want to sweat while they were sleeping either.

Direct mechanical refrigeration air-cooling systems—the immediate forerunners of modern air-conditioning—debuted in major New York buildings right after the turn of the twentieth century. For example, the new New York Stock Exchange building that debuted in 1901 was cooled without ice. These systems were commissioned by the architects who designed the buildings.[69] Starting in 1907, the Kroeschell Brothers Ice Machine Company began building mechanical refrigeration–based air-conditioning systems for other people's buildings. Most Americans didn't encounter air-conditioning until it became common in large urban movie theaters during the 1920s. Air-conditioning did not become common in ordinary houses until after 1930 because large mechanical systems especially designed for large buildings could not be shrunk down into window-sized units until after that date.[70] Therefore, overheated families that had not bought their first electric household refrigerators by that time undoubtedly bought ice from whatever source they could find.

While all these uses for ice are very different, they follow a similar pattern. The late nineteenth century saw a marked increase in the amount of ice available, making it easier to experiment with new uses. The growth of artificial ice making also contributed to the price dropping. That meant that when anybody used ice, they could use more of it, or at least worry less about much of it melting. Ice, this substance that was once common in winter but extraordinary in summer, became common everywhere at any time. Americans of all social classes made it an important part of their lives.

Technology creates incentives for economies to grow. Not only did improvements in harvesting, storing, and supplying ice make it possible for perishable food producers to grow more crops for Americans to consume, but they also created whole new industries by leading people to develop new uses for ice, like air-conditioning or ice-skating. Eventually, mechanical ice making became technologically sophisticated enough that it could expand into people's homes, which gave agricultural producers further incentives to create more perishable food, since people could buy more without having it go to waste. Over time, these electric household refrigerators would replace the entire ice industry, including both natural ice and mechanical ice (as it was first produced in the late nineteenth century) alike.

The appearance and growth of cheap and efficient electric household refrigerators would change the way we obtain ice forever. From one part of the kitchen to another is pretty much the shortest cold chain imaginable, but the importance of homemade ice cubes goes beyond that. Here for the first time was ice on demand, just as much as you needed and guaranteed to be as safe as the water you already drank every day. No contracts with an ice company. No ice deliveryman barging through your kitchen. Obtaining ice became simplicity itself. Pour water in a tray, wait a few hours, and squeeze out all the ice cubes you might need. While there is still an industry that supplies bags of ice for parties, that business does not operate in at all the same way as the ice producers of an earlier industry once did.

Yet the iceman persisted for longer than one might imagine, the same way that natural ice persisted even after warehouse-sized mechanical ice machines became cost-effective. The reasons for this depended on a number of factors. These include technical problems with early refrigerators, fears about the safety of such machines, and the availability of the electric current needed to run them. With strengths and weaknesses for both kinds of refrigeration during this period, a close examination of the interplay between the ice industry and the new electric household refrigeration industry offers the best window into understanding why ice persisted as an industry well into the twentieth century.

5 How Household Refrigerators Changed the Ice Market Forever

BY THE TURN OF THE TWENTIETH CENTURY, the ice industry had become an important part of American life. "America may well be proud of the ice industry, and may well claim its parentage," wrote the president of the Knickerbocker Ice Company (then the largest such firm in the United States) in 1895. "It brings comfort to the afflicted, it puts sweetness and purity in the place of decay, and by wasting it gives up its own life to save lives greater and more valuable. . . . It is a productive industry in the fullest sense."[1] The success of the American ice industry played an important part in improving the standard of living of people from all walks of life, just as the technologically sophisticated and highly refrigerated American perishable food provisioning system does today.

The cold chains that brought ice to the masses were one of many commodity chains that help explain how goods of all kinds are produced, distributed, and consumed. Culinary historians have focused on the history of food chains because access to any kind of food has always been affected by how easy or difficult that food was to obtain and preserve. Ice during the late nineteenth century was a particularly interesting commodity chain because its existence was so fleeting. Since melting is the way that ice makes and keeps anything cold, its fleeting nature is why ice required a vast industry to supply it in the first place. This system was so good at supplying ice that its low price served as a disincentive for most people and some companies to invest in mechanical refrigeration. But because ice melted, that system could only do so much.

The success of early mechanical refrigerators sprang directly from the limitations of the ice industry. Starting with the first unsuccessful household refrigerators, aspiring refrigerator manufacturers marketed their products as alternatives to ice delivery. The fact that consumers were interested in refrigeration at all derived from the market for ice which the ice industry had created decades earlier. While it is easy to see ice and refrigeration as totally different products, electric household refrigerators were really just another in a long line of improvements that made refrigeration in the kitchen cheaper and more efficient. In this way, ice set a precedent for the structure of the elaborate cold chains of today.

The entire transition from ice to mechanical refrigeration in America's homes took about forty years, from about 1920 to 1960. The fact that household refrigerators could essentially eradicate the American ice industry seems entirely logical. After all, the ability to make ice in your own kitchen shortened the cold chain for ice to essentially nothing. All it cost to make that ice was the cost of electricity and the cost of water. When consumers could make ice on demand, they were never left with food that might spoil for lack of an ice delivery. Similarly, you seldom had to buy more than you needed, except perhaps if you were throwing a party.

Given the many advantages of electric household refrigerators, one might wonder why this transition took so long. In the early years of refrigerator manufacture, these machines had many problems, including refrigerant leaks, corrosion of metals, vibrations, noise from the moving parts, and noise from the electric motor (which disturbed consumers when they were going about their daily business and even when sleeping). Other problems included the interruption of operation, lubrication troubles, and thermostat troubles.[2]

That changed only when large manufacturing firms such as General Motors and General Electric brought their research labs into the refrigerator business. These problems were only solved through systematic experimentation and, as with the ice industry, a certain amount of trial and error. While these problems were largely corrected by 1930, after that refrigerator buyers still needed enough money to afford an appliance and access to electricity to run it. Some rural sections of the country lacked a reliable electric grid even after World War II. Nevertheless, once the effectiveness of refrigerators improved and their prices dropped, the ice industry (in its nineteenth-century form) gradually disappeared.

Early refrigerator executives (many of whom were associated with the automobile industry) recognized that for refrigerators to succeed they had to be both reliable and cheap. Only when they achieved both these objectives simultaneously did the household refrigerator begin to assume its now-ubiquitous place in the American kitchen. General Electric (GE) began working on developing a reliable household refrigerator as far back as 1911. The earliest prototype was called the Audiffren, and it was based on an invention developed by a French monk, Abbe Audiffren. GE could only do this under license for the company that held the American rights for the design. Like so many other early mechanical refrigerators, the Audiffren failed because of its many operational flaws. Nevertheless, in 1917 GE began research to see whether it could create a mechanical household refrigerator based on the Audiffren design.[3]

This early research led directly to the development in 1927 of the "Monitor Top," so named because of the resemblance of the machinery casing on the top to the upper section of a Civil War battleship. GE's Monitor Top introduced countless Americans to all the advantages of electric household refrigeration. It became the first serious threat to the hegemony of the ice industry and should be considered the first modern electric household refrigerator. The first Monitor Tops had a storage capacity of between five and seven cubic feet, along with an ice cube tray. The boxes where the food went were also elevated off the floor by eight-inch legs, which gave it more of a resemblance to a piece of furniture than the box-shaped refrigerators of today.[4]

Yet other traits that made the Monitor Top different from the competition at the time echo in refrigerators even now. The most important of these traits was its all-in-one design. The DOMELRE and many of GE's other competitors at that time were nothing but small machines intended to be installed in existing iceboxes. GE quickly decided against going that route because so many iceboxes in people's homes were less than dependable.[5] Kelvinator, another competitor, had required buyers to cut a hole in their kitchen floors so that a belt could connect the container to the condensing machinery, which went in the basement.[6] One of the Monitor Top's most outstanding features was the hermetically sealed machinery, bathed in oil in order to reduce noise.

Independent of operating concerns, the greatest problem that electric household refrigerators faced in the quest for acceptance was the cost of the current needed to run them. "It uses very little current and no special wiring to set it up," explained GE in their first advertisement for the final version of their refrigerator. All you had to do was plug it into the wall.[7] Fifty kilowatt-hours per month powered the Monitor Top, which was half the power required to operate competing refrigerators.[8] Equally important, once GE got its assembly line up and run-

First electric refrigerator produced in a large eastern plant, 1928. Library of Congress

ning, it could begin to sell the Monitor Top for just $525, a price that middle-class consumers could afford.[9]

The technological sophistication of this refrigerator, combined with its low price, made the Monitor Top an instant success. "It had taken more than fifteen years to develop and perfect it," GE later bragged. "[GE research engineers] would not announce it nor permit it to be offered for sale until they were certain it would prove to be the most practical and the most economical mechanical refrigerator ever produced."[10] Another GE ad claimed that the Monitor Top was "so efficient, so simplified, that it literally could be installed and forgotten."[11] Spurred on by a huge advertising campaign, GE sold 50,000 Monitor Tops by 1929. By 1931, total sales reached one million.[12]

With mass production and mass consumption of refrigerators came consolidation of the industry. In 1931, there were seventy-five different refrigerator manufacturers in the United States.[13] By the start of World War II, the industry was dominated by just four firms because of the barriers to entry that mass production imposed. In order to facilitate mass consumption, consumers could purchase refrigerators in more places than ever as the 1930s began. While the refrigeration industry had once depended on electrical utilities to sell their wares, by the 1930s you could buy refrigerators at automotive supply stores.[14] When Sears introduced its well-received Coldspot refrigerator in 1928, it became possible to buy these appliances by mail order. Greater availability was just one reason that sales grew over the course of the decade, despite the Depression.

When GE first introduced the Monitor Top, it was not really competing against other refrigerators. People who had bought one of the pioneering appliances from the years 1915 to 1925 would have been reluctant to give those up to get a whole new refrigerator so soon. Instead, GE had to convince consumers who depended on ice to make the Monitor Top their first electric household refrigerator. "Mrs. Smith," went a sales pitch script for Monitor Top salesmen in a guide published around 1928, "I am not here to sell you anything today—or tomorrow. But some day I am sure you will want to enjoy the benefits of electric refrigeration. There are many users in your city and I am sure that some of them are your friends." In fact, salesmen were specifically instructed not to knock the ice industry or even competing refrigerators.[15]

Only when the customer decided that they wanted to buy an electric refrigerator would the salesman pitch the unique advantages of the Monitor Top. These advantages included the "absolute freedom from attention" (meaning it ran by itself unattended), its quietness, its long life (since the moving parts were hermetically sealed in a bath of oil), its low operating costs (since it used relatively little electricity), and its relative portability if you were to move to a different house. This pitch was usually given at home, in the hopes that the salesman could reach the husband and wife at the same time. "We will assume that most men will buy a 'General Electric' [refrigerator] if their wives request it," the salesman's manual explained. "It is easy to see then that the desire to own the 'General Electric' is usually created with the wife in the home."[16] After all, she was the one who had to deal with all the hassles produced by buying ice from the icemen—everything from deciding on the terms of a contract to cleaning up after the iceman's dirty boots after he left the kitchen. No wonder refrigerator salesmen targeted her.

The differences between electric refrigeration and ice also explain why GE touted the sanitary value of the clean, dry cold that the Monitor Top produced. "The dryness of the atmosphere prevents the multiplication of the bacteria which every food contains, and which thrive in dampness," the company wrote. "The compartments, which are lined with spotless, white, vitreous enamel—which does not darken with age—can be cleaned with little trouble."[17] While icebox owners no longer had to deal with the sediment that often permeated natural ice, manufactured ice still produced a wet cold. Perhaps most importantly, the cold that ice produced wasn't particularly cold. Ice couldn't go below thirty-three degrees unless you added salt to it (which, of course, could have corroded the metal in early refrigerators). Since every early household refrigerator had a freezer compartment, it could get much colder. This meant that it could keep food from perishing longer than ice ever could.

Because ice was both wet and dirty, iceboxes had to be cleaned thoroughly and often—especially wooden ones. The Monitor Top, on the other hand, was made of white enamel. It could be cleaned with a damp cloth. "Ample space is allowed between the chilling unit and the sides of the cabinet, in order that the remotest corner may be cleaned easily and well," the company explained.[18] A later GE pamphlet made this unflattering comparison more explicitly, noting that the Monitor Top "has no exposed wood to absorb food odors. It is supremely easy to clean." GE also bragged that the circulation of air kept dust from the top of the container.[19] Along the same lines, one of the company's first advertisements read, "All the models are up on legs. This makes it so easy to keep the floor underneath clean."[20] All these small advantages were supposed to add up to one superior refrigerator, at a time when people still needed some prompting to make the leap to a new kind of electric appliance.

While we tend to take access to ice cubes for granted today, the ability of the Monitor Top to make its own ice was, in fact, one of its strongest selling points in the early days of the household refrigerator. That ice, explained GE, came "in small, clear cubes from your drinking water. . . . These ice cubes have the advantage of absolute purity. They are made in the home and never touched, except by the person who uses them." This was a jab at the iceman for dragging large blocks through dirty streets. Indeed, GE noted that its ice cubes came "in a most convenient and dainty form."[21] The Monitor Top even came with a "quick freezing feature" (basically slots for ice cube trays very close to the machinery) so that customers could get cubes before the rest of the unit had completely cooled down to its normal temperature.[22]

To cement its position at the top of the refrigerator market, GE put a three-year guarantee on its Monitor Tops for the first time in 1930.[23] Whatever concerns that had once existed about the reliability of household refrigerators were well on their way to disappearing by then. No wonder GE bragged in its salesman's handbook that, "in keeping with the speed of material progress in this country and the world today, we may confidently expect that the present generation will live to witness and enjoy a very general use of electric refrigeration for a wide variety of purposes."[24] Unspoken in that very accurate prediction was that the ice industry would not be around much longer.

Because of the success of this massive promotional effort and the superiority of this appliance, the Monitor Top was at the forefront of a huge boom in electric household refrigerator sales overall during the late 1920s and early 1930s. From 1,680,000 refrigerators in households nationwide at the end of 1929, the total number of refrigerators in America jumped to 7,250,000 at the beginning of 1936—despite the depths of the Depression. During the same general period, the average price of a refrigerator dropped by about half, and the cost of the electricity to run it dropped from somewhere between a quarter and a half. These newer refrigerators were so much better than the older ones that consumers began turning in their first refrigerators to buy new ones, even if their old ones worked fine.[25] In short, around 1930, the household refrigerator had left the experimental stage and became a permanent fixture in the American household.

In the meantime, the ice industry was "more or less in a state of chaos," as ice producer Halbert Hall explained at a meeting of his peers in 1927. The reasons he cited for this chaos were "the keen competition within its ranks, the household machine, the weather conditions of 1926 and 1927, together with the lack of a set-forward program to stabilize the industry." In an effort to exhort his fellow ice manufacturers to act, he cited the example of an earlier, related industry that had been thriving until relatively recently. "You must all realize that prior to 1900 the natural ice industry in this country was a very healthy enterprise," he told them. "What I am trying to point out to you is that with the change of time and conditions there is that ever present competitor, Progress, to contend with."[26] Hall meant this as a cautionary tale for all the ice manufacturers in the room who thought that their business could survive this chaos intact.

Unfortunately for the ice industry, it took a considerable amount of time for its leaders to take this threat to their very existence seriously. Instead, the ice industry's initial response to the threat that the electric household refrigerator posed was to ignore it. This made sense in the early 1920s, when refrigerators didn't work very well. However, as electric refrigerators got better and became more popular, most of the people who produced and sold ice for refrigeration did little to address this growing threat. "I reiterate my belief that the domestic mechanical refrigerator will play an important part in domestic refrigeration," explained John Nickerson, the publisher of the industry's largest trade journal in

November 1926, "but it will never be the dominating factor in the industry."[27] In 1930, total electric refrigerator sales exceeded icebox sales for the first time—and never looked back.[28]

Despite losing its place as the dominant source of household refrigeration in the American market, the ice industry still grew during the 1920s because prosperity drove up demand. So did the constant call of refrigerator manufacturers telling people how their safety demanded that they keep their perishable food cold. Those who could not refrigerate such foods bought ice instead. As a result of these factors, demand for ice in the United States did not collapse until 1932.[29] At that point, any consumer who was able to purchase and operate a new refrigerator generally did so. The remnants of the ice market persisted only because the Depression kept some people from being able to purchase a refrigerator and lack of electricity (particularly in rural areas) prevented others from being able to operate one.

With the large ice manufacturers unable to recognize the devastating impact of the household refrigerator on their livelihoods until it was too late, smaller producers and ice delivery companies took the offensive in defense of their product. To do this, they found the few remaining problems that electric refrigeration faced and did their best to magnify them. "Do You Know," asked one of the points on a list of twelve "Cold Facts" written by a Kansas City, Missouri, ice and cold storage firm and quoted by *Ice and Refrigeration*, "that the annual cost of electric current alone, for the household mechanical refrigerator, will pay the ice bill for three average families for one year[?]"[30] The efficiency of the Monitor Top and later refrigerators eliminated this problem soon after.

Another line of attack this company adopted played on the reliability of electric household refrigerators to that point in time. "Do You Know," they asked, "that ice is ready for use and performs its service the very moment it is placed in your [icebox]; that you can fill your ice chamber and be gone for a day or two and know that your food is absolutely protected while you are away; no fuses to blow, belts to break, no parts to oil and it is unaffected by electrical troubles?"[31] Another campaign from a different Kansas City ice company which appeared later that same year summarized this line of argument with the slogan, "Ice does not get out of order."[32] But as with the decreasing electrical costs and as suggested by GE's eventual willingness to offer a three-year guarantee, the reliability of refrigerators improved quickly enough that this line of argument did no lasting damage.

While *Ice and Refrigeration* reported that the many campaigns like these were "all . . . very effective," some of the ice industry's more private efforts probably had the greatest impact on slowing the spread of electric household refrigerators. One refrigerator company executive later remembered how the ice companies got local ordinances passed to impede their business: "We were required to employ union men instead of our own to do the wiring. We had to employ licensed plumbers to install the machines instead of our own skilled and experienced men."[33] Had the electric refrigerator been developed by the ice industry, this might have played out differently, but the appliance manufacturers were outsiders—usually connected to the automobile industry. They not only managed to reconcile the technological requirements of home refrigeration but also understood the ethos of mass production which made the mass adoption of their appliance possible.

The ice industry also undertook more positive improvements in order to deal with the threat posed by home refrigeration. For example, they greatly improved service. One sign of these improvements was the introduction of standardized ice blocks and scoring machines (to make breaking those blocks easier) during the late 1920s. This meant that ice companies could divide their blocks into the size of individual customers' orders before they ever left the plant. This made it impossible for shady ice deliverymen to cheat their customers by short-weighting them. It also sped delivery by shortening the amount of time that an iceman had to spend with each customer.[34] Another development that dates from this era was the "crusher slinger" or "crusher blower" machine—a huge contraption that could turn a three-hundred-pound cake into cubes in about ten seconds. This invention made it possible for ice plants to start producing ice vending machines, since icemen no longer had to break blocks up for customers in order for them to get ice.[35]

In order to ensure that its customers could preserve more of the ice they got, the ice industry began to enter the icebox sales field. Frederic Tudor had done precisely this in the early days of ice harvesting, but ice companies grew complacent by the late nineteenth century. Bad iceboxes meant more melting, and faster melting meant buying more ice to replace what melted. However, now that electric refrigerators challenged the ice companies' stranglehold on household refrigeration, building a better icebox became a marketing necessity. Improvements in design, insulation, and even ice trays meant that visits from the iceman could be spaced out from between one and two days to intervals as long as every four to seven days.[36]

The ice industry also did its best to expand into new markets. Its greatest success came in comfort cooling—air-conditioning in the days before both building-

Ice and carry station, created so that customers could pick up ice rather than wait for a deliveryman. In order to compete against abundant ice produced by household refrigerators, ice companies had to make their product more convenient. Nickerson and Collins Refrigeration Photographs, Archives Center, National Museum of American History, Smithsonian Institution

sized air-conditioning systems and household air-conditioning units had become cheap and efficient. Articles touting the many benefits of ice cooling began to appear in 1931, just about the time that ice manufacturers were waking up to the threat that household refrigerators posed to their business model. Even though the ice in ice-cooling systems had to be constantly replaced, the price of ice had become so low (with demand dropping) that these kinds of systems were affordable for the time being. Building contracts from the early 1930s suggest that commercial installations during this era generally required six tons of ice per day.[37] However, this temporary boost to the ice market could only help a dying industry stay afloat for so long.

The initial success of the Monitor Top did not prevent GE (or its competitors, for that matter) from making refrigerators even better than they already were.

From the moment of its initial launch in 1927, GE introduced different models of the Monitor Top intended for different kitchens. The more space (and presumably the more people) you had in your house, the bigger your new refrigerator could be. GE also produced commercial refrigerators that made use of the parts and features developed for the Monitor Top. These refrigerators were installed in places like stores, restaurants, meat markets, and hospitals.[38] Like consumers, these institutions, once they bought an electric refrigerator, would no longer need to purchase ice.

GE also added more features to Monitor Top models over time.[39] In 1931, for example, GE advertised new features such as "sliding shelves, acid-resisting porcelain interiors and the Sanitary Super-Freezer."[40] GE also made small design modifications each year so that new buyers would be able to feel that they had purchased a state-of-the-art model. More importantly, GE improved the performance of the Monitor Top by making technological changes. In 1931, it introduced Thermocraft insulation. This not only reduced the weight of the refrigerator but also did a better job of preventing heat from getting into the cabinets. This insulation was so successful that GE licensed it to their competitors for years after it was first introduced.[41] Among the other improvements were better lubrication of the moving parts and even quieter operation.[42]

Perhaps the most important change that GE made to its refrigerators during these years was the introduction of a stripped-down model in order to encourage first-time buyers to jump into the market. The first Sears Coldspot refrigerators had been technical disasters, built with poor quality control. Much of the Coldspot's appeal was based on its lower sticker price. Sears hired industrial designer Raymond Loewy to redesign the Coldspot three times over the course of the 1930s. The result was a bigger refrigerator that looked more modern, but it did not have as many features as its more expensive competitors. As a result, Sears went from the eleventh-largest refrigerator manufacturer in America to the second-largest one by 1939.[43]

In order to get a version of the Monitor Top design down into the Coldspot's price range, GE took features away rather than adding them. They did so with the explicit encouragement of the Roosevelt administration, which had begun financing refrigerator purchases after the Tennessee Valley Authority brought electricity to a large swath of the country which didn't have it before. In 1934, GE introduced a refrigerator for $74.50—$20 less than its lowest competitor.[44] Whatever the company lost in markups they more than made up for by encouraging higher total sales.

Because many other refrigerator manufacturers followed this same strategy, household refrigeration became gradually accessible to just about everyone in America. By 1939, these stripped boxes retailed between $135 and $150, but at six cubic feet they were larger than similar GE and Sears models. That appealed to cost-conscious consumers with families. This, along with a huge increase in people coming back to buy their second refrigerator, led to a spike in refrigerator sales that year. Besides being lower in price than earlier refrigerator models, these new refrigerators were larger, were better looking, and lasted longer too.[45]

Faced with stiffer competition, GE introduced its first new refrigerator since the Monitor Top in 1934. Instead of having the machinery on top, the machinery on this model was located under the food compartment.[46] This explains the name "Flat Top." The Flat Top also had a decorative panel on the front, like so many other objects during this design-conscious era. The Flat Tops were even manufactured differently, with each side made as one piece and fashioned to a skeletal frame. This helped GE and other manufacturers who copied this method keep the price down. GE stopped manufacturing the Monitor Top in 1937 mostly for aesthetic reasons. The Flat Top had the same essential guts as the Monitor Top, but it sold better.[47]

Like the Monitor Top, refrigerators of all makes and price ranges were improving throughout the 1930s. Refrigerators made in 1940 used about half the power of models ten years old at that time. The newer refrigerators also ran quieter than the old ones. "The noisiest unit built today is quieter than the quietest built five years ago," concluded a talk published in *Refrigerating Engineering* in 1937.[48] New amenities had also appeared in refrigerators during the 1930s, such as meat and vegetable compartments and compartments that kept particular foods under high humidity. This was also the decade in which manufacturers introduced a light that went on inside the refrigerator when you opened the door. Like the Monitor Top, other refrigerators were designed better and priced lower than ever before.[49]

Less obvious improvements in refrigerators also improved sales. New kinds of seals eliminated refrigerant leaks of all kinds. Improvements in electric motors made them less noisy, lighter, and less expensive. Temperature control became practically flawless by 1930. Manufacturers replaced wooden refrigerator cabinets with steel containers, many of which were coated with porcelain enamel. Energy efficiency in refrigerators also improved.[50] Tests performed on a late-model Monitor Top in 1935–36 showed that it used 20.3 kilowatt-hours of electricity per month. Refrigerators in the late twentieth century (because of their larger size, especially their large freezers) used between sixty and one hundred kilowatt-hours per month.[51]

Refrigerator manufacturers even managed to improve the process of obtaining ice cubes. The earliest ice cube trays consisted of removable brass grids that fit into trays of heavy, nickel-plated copper. Consumers had to run the frozen trays under running water in order to get the cubes out of them. Lloyd Copeman designed the first flexible rubber ice cube trays, which appeared in the late 1920s. Just a twist and the ice would fall out. In 1933, Guy L. Tinkham, a vice president at a household appliance company, introduced flexible metal trays.[52] "New refrigerators yield their ice cubes without a struggle and without the wasteful melting of the cubes," explained a story in the June 1940 issue of *Good Housekeeping*. "More ice cubes and faster freezing assure a plentiful supply of ice cubes whenever you need them."[53] A typical refrigerator in 1937 could produce seventy pounds of ice cubes between 8:00 a.m. and midnight every day.[54] With such commonplace efficiency, ice cubes were no longer a selling point for new household refrigerators by the late 1930s.

Despite all these improvements in refrigerators, the ice industry did not completely disappear when electric household refrigerators became popular during the Depression. Only one in nine American farms had electricity by the middle of the 1930s.[55] These Americans still needed ice, assuming they had access to refrigeration at all. After all, what good was an electric refrigerator to a household that could not purchase electricity because it was too expensive to bring it to them? Other Americans still opted for ice by choice. As late as 1935, 69 percent of homes in the quintessentially middle-American town of Muncie, Indiana, still used ice.[56] In 1937, Americans still bought $9,981,000 worth of iceboxes.[57] World War II was the event that forever changed the way Americans used ice.

After its success with encouraging electric refrigeration through the Tennessee Valley Authority, the Roosevelt administration began to subsidize refrigerator purchases with the National Housing Act of 1935 and through similar programs. They considered this to be an essential part of modernizing the American home.[58] It is no coincidence that the last households to get refrigeration were among the last ones to get electricity. The Rural Electrification Administration (REA), a New Deal agency formed in 1935, was still introducing some rural areas of the United States to electricity in the late 1940s. At that time, consumers in these areas faced the same icebox versus refrigerator decision that much of the rest of America had made twenty years before.

A rural Georgia housewife uses her refrigerator, paid for in part by the Rural Electrification Association, 1941. National Archives via Wikimedia Commons

When they could afford it but before the wartime production needs made refrigerators unavailable, countless consumers bought their first (and by then sometimes their second) household refrigerator. At the start of World War II, just under half of American homes owned a refrigerator.[59] That number grew during the war because the War Production Board did not stop electric household refrigerator production so that those plants could be converted to wartime needs until April 30, 1942, and they let manufacturers ramp up production during the weeks before that deadline.[60] As a result, 85 percent of American homes had a refrigerator in them by 1944.[61]

A general push for rural electrification in order to aid food production during the war led to a greater incitement to continue that process after the war ended. People who had gotten used to the many benefits of refrigeration (especially for employing labor-saving devices that boosted production) wanted more of these kinds of benefits. By the 1950s, just about every home in America which was wired for electricity had a fridge in it.[62] Beyond the many benefits of mechanical refrigeration, this meant that every one of these households no longer needed to purchase ice.

The government still took the possibility of families electing to purchase an icebox seriously after the war, despite the enormous penetration of electric refrigerators into the rest of the country. "If the refrigerator is to be iced, the ice capacity needed may be dependent upon the frequency of delivery," explained a fact sheet that REA agents used in the field. "Regularity and frequency of service are important." The agency also noted that iceboxes were cheaper than household refrigerators, but then cautioned that people interested in buying iceboxes still needed to look at the cost of operation.[63] This suggests that while there were once more ice delivery services in urban areas, the ones that survived household refrigerators were located in rural areas where refrigerators could not run before the REA brought current there. Some of those services located in far northern latitudes may even have still depended on natural ice.

Of course, that assumes that people had access to ice delivery. Many farmers didn't even have iceboxes before the REA came along because it was so hard to deliver ice in the countryside. Rural delivery wasn't economical because the nearest ice plants were too far away. All the ice would have melted before it got there, assuming that anyone was willing to deliver it. That would explain why the agency deliberately used the possibility of introducing electric household refrigerators to areas that didn't have them as a major selling point for their services. "I don't need to tell you that you are getting a real buy," explained a script from 1941 for the men who delivered and installed people's first refrigerators. "You know that yourself. But don't forget that this was made possible through cooperative group action with the help of REA. That's the way we get electricity at cost. That's the way you are getting a good refrigerator at a price with most of the distribution costs eliminated."[64] Without this kind of government help, the traditional ice industry might have survived a little longer, at least in the places that it actually served.

In 1944, there were seven and a half million farm dwellings in the United States. Only three million of them were wired for electricity.[65] People who lived in areas without electricity still had no choice but to use ice for their refrigeration needs. A US Department of Agriculture booklet from around 1950 acknowledged the possibility that some people may still want to buy an "ice refrigerator," but after mentioning the importance of regular delivery, it then ignores icebox issues for nearly the remainder of the publication. Nevertheless, some companies continued to sell iceboxes even at this late date.[66]

Rural farm families that did not have access to ice were particularly grateful when they could finally buy their first refrigerator. "Before electricity came to us, I had to store milk, vegetable and butter either on the screened-in back porch, or in the basement," wrote a farm housewife from Bolivar, Missouri, in 1952, having gotten electricity just three years earlier. "Now that electricity has come to our farm, all this has changed. . . . My electric refrigerator is not only a modern place to store food, but there are also so many tasty desserts that can be made putting it into use. In fact, one might think that my boys think more of the refrigerator than their mother—they always head for it first when they arrive home from school."[67] How could iceboxes ever compete with this? Electricity allowed farm families like this one to skip sixty years of refrigeration history and go straight from using root cellars to electric household refrigeration.

In rural areas that still clung to ice delivery after 1950, the scale needed to run many of those companies would have become impossible as people switched to electric household refrigerators. In those instances it seems likely that ice delivery services failed before their iceboxes did. Their customers would have switched the moment they got a chance to purchase their first refrigerators. By then refrigerators wouldn't have seemed quite so new anymore. Consumers would have seen them in magazines or heard about them on the radio. Therefore, these customers would have known all the advantages of the jump they were about to make. Furthermore, by that time all the disadvantages associated with early refrigerators would have disappeared.

"It was but three or four decades ago that every farm had its ice house," explained an article in the magazine *Boating* from 1960. "Stores and homes received a daily ice delivery. . . . Now the various accouterments of the 'Ice Age' are gone. The cost of operating a home type electric refrigerator is far less than supplying the older type with ice. Inconveniences due to the use of ice have been eliminated. With two great factors, convenience and economy, working against him, it is little wonder that the Ice Man has joined the ranks of 'Vanishing Americans.'"[68] The last ice deliverymen gradually disappeared over the course of the 1950s because refrigerators became so cheap and so ubiquitous that there was simply no way left for an ice delivery company to make a profit. What made them ubiquitous was the availability of electricity practically everywhere in the United States. By 1960, both the electrical grid and the modern cold chain were essentially complete.

In the early 1950s, total ice production in the United States was close to what it had been fifty years earlier, but consumers bought almost none of it. The ice industry survived in commercial markets. Railroads still needed ice to ship produce. Markets and cafeterias still needed it to preserve and present their offerings. Yet mechanical refrigeration was still changing the way that we consumed perishable food during the post–World War II period. Grocery stores switched over to mechanically refrigerated storage units and stopped buying ice. New vending machines could keep soda cold enough that purchasers could drink it straight out of the bottle or can without any ice cubes.[69]

Starting in the 1960s, the ice business reinvented its product as a kind of convenience food. The first "tube ice" machine appeared in 1937, but it didn't take off until the 1950s. These machines froze ice in tubes and cut it into easy-to-bag cubes. They were smaller, were cheaper to produce, and used less electricity than their larger commercial competitors. By the 1970s, the ice machine business experienced double-digit growth every year. Since making your own ice took way too much time and space for consumers planning a party or barbecue, they visited the grocery or liquor store to buy cheap bags of all the ice they needed, with no deliveryman necessary. In 2001, the largest ice cube manufacturer in the United States had revenues of 250 million dollars.[70]

What has changed most about ice between the advent of household refrigerators and today is our access to it. While it once paid to manufacture ice in large plants and deliver it to people's doors, we can now reliably make all the ice we need for most occasions in our homes with little trouble. The cold chains—long strings of various technologies of varying complexities needed to get ice from the point of production to the point of its consumption—are no longer necessary. Indeed, since most of us have faith in the purity of our tap water, the ice we make ourselves is a lot less scary than all the packaged foods that we consume, full of ingredients whose names we do not really understand.

Ice production—whether through harvesting or manufacturing—was once a technology that few people understood. That explains why the print media in the nineteenth and early twentieth centuries devoted so many words to explaining from where ice came. Now we know exactly where ice comes from. Except for special occasions, ice comes from our freezers. We no longer require plows, or ice picks, or deliverymen in order to obtain it. While we may not entirely understand how our own refrigerator works, the short lengths of our cold chains obviate any worries we may have about the purity of our ice. This may be the primary advantage of today's ice over the ice that people consumed a hundred years ago. Thanks to our refrigerators, ice is always as good as the water from which it comes.

Between 1880 and the early 1930s, ice went from being the sole source of refrigeration in the United States to being simply a source of refreshment. Certainly ice still plays important roles in commercial food chains today, like keeping the skins of fish moist or refreshing lettuce in an ice bath, but mechanical refrigeration can do almost everything that ice once did with respect to keeping food fresh, and better than ice ever could. Many Americans still put ice in their drinks, even though it inevitably dilutes whatever we happen to be drinking.

It once took a vast array of technologies to harvest, transport, store, and use ice. By shortening the cold chain, household refrigerators made most of those technologies unnecessary. Now getting ice is so simple that most Americans take it entirely for granted. Yet ice can be precious for life and health in lesser-developed countries throughout the world. These nations, like many developed nations in Europe, skipped the ice plant and delivery stage of cold chain development and have been getting household refrigerators soon after electricity arrives, just like the rural folks in the post–World War II United States. This means that how we used to get ice was a uniquely American experience. The evolution of the system by which ice got to us, on the other hand, bears similarities to systems that developed to bring us other comforts over the course of the twentieth century.

CONCLUSION. The Inevitability of Melting

ICE MELTS. That may seem so obvious that it doesn't need mentioning, but it is the most important fact to keep in mind if you want to understand the entire system of technologies that brought ice to consumers during the late nineteenth and early twentieth centuries. Every technology that the natural ice industry utilized had to take into account the fact of melting, because the more ice melted, the less of it anyone would have to use or sell. Still, as late as 1883, ice harvesters lost 25 percent of their product just waiting for the summer to roll around and another 25 percent during delivery.[1] Even when mechanical refrigeration became technologically viable, that technology was not portable until many years later. Therefore, even ice manufacturers lost their product as a result of melting during transportation or delivery. The easiest way to make melting less important was to produce more ice.

The American ice industry of the late nineteenth and early twentieth centuries managed to increase the total ice supply by shortening their cold chains, rather than expanding them as Tudor did. They did this primarily by manufacturing ice on a large scale. With a shorter distance to travel, the supply of ice increased because it had less time to melt. The great advantage that mechanical ice had over its natural counterpart was that it could defy geography and the climate conditions that came with it. Warm areas could now have as much ice as they wanted. In the American South, where natural ice was no longer economically viable after 1880, this became the only option that ice purchasers had. Therefore, they flocked to it even before the ice that these machines made was any good.

It took less than a century for the ice industry to create systems efficient enough for melting to become a manageable fact of life rather than an insurmountable impediment to its existence. Besides shortening the cold chain, the key to success was to get ice to melt when and where it was most needed. A vast variety of technologies implemented gradually over the first hundred years of the industry's existence assured that this would happen. Some of these ice-related technologies were very complex. Some of them were very simple. Some were expensive. Some were cheap. Some were long-lasting. Others were improved on so much so that earlier technologies that had once been crucial for the ice industry's existence became obsolete long before ice itself did. As the technologies of ice harvesting and production improved, so did the quality of the ice that this industry provided.

Because ice was the only way that people had to control the cold in their everyday lives, this made it worth preserving. Every technology that developed in order to make the ice production and distribution system more efficient was at some level an effort to compensate for the fact that ice melts. Consider, for example, the three potential museum exhibits from the introduction to this book. An ice auger that measured the thickness of natural ice assured harvesters that the ice was thick enough that it could make it through the winter and out on delivery without melting completely. The outside ice protected the inside ice (as it did in Tudor's day). Therefore, the thicker the ice, the more that would survive the entire cold chain, which meant more to sell.

The coupon book, which prevented icemen from cheating customers and their employers, was designed to record the existence of a highly fleeting substance. When icemen were forced to keep records of how much ice they had, how much they sold, and how much their customers received, it became extremely difficult to sell a little on the side and chalk it up to melting. Indeed, the better the record keeping, the easier it would have been for ice companies to know how much ice they were losing to melting every day of summer. Just because ice melts doesn't mean that the product these firms sold couldn't be tracked like everything else. By 1920, artificial ice manufacturers had started producing standard four-hundred-pound blocks of ice, but they actually weighed 416 pounds. This was to compensate for the inevitable melting, allowing delivery companies to track their business more easily.[2]

Even the appeal of the first artificial ice machines derived in part from the tendency of ice to melt. Instead of collecting ice from a single distant point and distributing it to multiple places throughout a city, ice machines placed at strategic locales throughout a city shortened the cold chain, thereby leaving more ice to sell. It was this innovation more than anything else that made the supply of ice essentially limitless in urban areas when ice manufacturers appeared. This, in turn, broke ice production away from the cycle of the seasons.

On the other hand, it had been easy to control local natural ice sources because there were only so many places near any city where it could be cut. Because ice melts, competition from outside their local areas was next to impossible. Even artificial ice producers could try to limit supply by not making as much, but in those instances they faced the problem of new actors entering the market. When ice machines became viable and relatively inexpensive, those trusts could not survive long. Ice production and distribution systems had many actors who stood to benefit from the many technologies that they could use in order to compete effectively.

No single person built these distinct but related ice production and distribution systems. Yet these were systems nonetheless. To borrow a distinction from the historian Thomas Hughes, these were vertical systems with various interconnected components all designed to bring a single product (be it ice or perishable food) from the point of its production to the point of its consumption. The other kind of system that Hughes considered were horizontal systems, like electrical power plants that were subject to central control.[3] Recognizing that ice required a system in order to make it a viable commodity, one can see that the fact that ice melts made it about the least horizontal system imaginable. By the time that any central figure could coordinate production at all their plants, it would melt. Besides that problem, anyone with a commercial interest in refrigeration could undermine attempts to limit supply by harvesting their own ice for use at the factory level or buying an ice machine of their own.

The story told here demonstrates the economic value of keeping these systems decentralized. Small but significant improvements like coupon books or new ways to pack ships were not the kinds of innovations that could be patented. Therefore, they spread quickly through the ice industry, making the whole system more efficient. Those kinds of innovations that could be patented, like a particular kind of ice machine, did not preclude other inventors from trying to create a more efficient ice machine through other means. “Why did Edison so often choose to work on systems?" asked Hughes in reference to his work on electrical systems. "If the inventor created only a component, he remained dependent on others to invent or supply other components. The inventor of a component could not have the control over innovation that Edison wanted."[4] The vast array of technologies required to produce, store, transport, and consume ice might be controlled by a single company in a single location for a short amount of time, but any attempt to bring those advantages to another city required every potential monopolist to start from scratch. Ice manufacturing had economies of scale, but that advantage disappeared once that plant dominated the area through which ice could travel before it melted.

In his book *American Genesis*, Hughes made an analogy that's also useful for better understanding the early ice production, storage, and distribution system. He borrows the term *reverse salient*, meaning a military front that develops unevenly, and applies it to technological development.[5] The ice industry in America, which depended on an enormous array of technologies operating independently in local markets all around the country, developed extraordinarily unevenly. The largest ice machines, for example, could easily produce more ice than could be sold or economically stored in most cities. They did so in large part because the means for its transportation—be it by land or water—could not be refrigerated itself until well into the twentieth century. Similarly, families without electric refrigerators (which included everybody before 1915 or so) had to buy much more ice than they needed because the technology to make ice had advanced so much faster than the technology to preserve it.

Contrasting the American ice industry with Henry Ford's system of mass production (which dates from the same general period) is a good way to understand how this system could have so many independent parts and still function effectively to meet the country's ever-expanding desire for ice. In order to maintain quality control for his Model T automobiles, Ford tried to bring every part of the production process in-house. This prevented hypothetical subcontractors from damaging his business by supplying substandard parts for which they could not be held responsible. Ice, on the other hand, was only as good as the water from which it came. Indeed, if consumers were not selling ice for human consumption, even the quality of the water did not really matter because cold was cold. What did matter was how much ice remained once it arrived where it was going. This gave every link in the cold chain incentive to preserve all the ice it could until it reached the consumer. Even then, businesses that sold ice to consumers had an interest in making sure that consumers could have as much ice around as possible, whenever they wanted it. Otherwise, they might have experimented with different systems or perhaps forgone refrigeration entirely.

The fact that ice melts is also what eventually inspired increased regulation over this particular industry. While other systems, particularly manufacturing systems, could produce as much as national and even international markets could absorb, the only way that ice manufacturing could prove anywhere near as profitable was to limit the free exchange of this commodity. Until mechanically refrigerated transport came along, the inherent geographical limits of ice distribution prevented any need for corporate capitalism to limit its own excesses. Since it could not travel far, all those little ice trusts were easy to create and maintain. Therefore, different levels of government, in the form of everything from antitrust actions to the establishment of municipal ice plants in cities across the United States, did their best to limit the local excesses of this industry.

The first hint of state interference with the ice industry came around 1900, when local governments began limiting the ability of ice companies to harvest from polluted waters. By the 1910s, ice was treated more like a utility than an industrial corporation because people in power were convinced that nobody could live without it and that it was the government's responsibility to make sure that they would not have to do so. For example, in 1925 the state of Oklahoma declared that ice manufacturing was a "public business" and that no potential ice manufacturer in that state could make ice without a license from that state's Corporation Commission. The New State Ice Company got a license, built a plant that cost $500,000, and began to sell ice. When a competitor began to build a plant without a license, the New State Ice Company sued. That case went all the way to the United States Supreme Court, which decided in favor of the plaintiff in 1932.

Justice Sutherland, who wrote the majority opinion in *New State Ice Company v. Liebmann*, called ice manufacturing "a business as essentially private in its nature as the business of the grocer, the dairyman, the butcher, the baker, the shoemaker, or the tailor." The advent of the electric household refrigerator eliminated the need to consider the ice industry a "public business" anymore. "It is common knowledge," Sutherland wrote, "that today, to say nothing of other means, wherever electricity or gas is available (and one or the other is available in practically every part of the country), any one for a comparatively moderate outlay may have set up in his kitchen an appliance by means of which he may manufacture ice for himself. Under such circumstances it hardly will do to say that people generally are at the mercy of the manufacturer, seller, and distributer of ice for ordinary needs."[6] The cold chain was never really controlled by a single company. There were too many different products (including ice) moving through it for one company to master all of it.

This is how the ice industry became a triumph of decentralized capitalism. A whole system came about out because of individual interest in making ice affordable to consumers, which in turn benefitted people all the way from harvesters and ice machine makers down to the people who used it and bought the food and drink that depended on ice for making their product into a mass-produced commodity. In this manner, the benefits of both production and consumption spread out among as many people as possible. No single inventor or entrepreneur made this situation possible. Instead, many inventors and ice entrepreneurs created a web of cold chains out of many different parts that worked toward a common goal.

Cheap, reliable mechanical refrigeration has now made nearly all the technologies that the ice industry depended on obsolete. Today, making ice is extraordinarily easy: find a cold environment (probably in your very own kitchen), insert water into a receptacle (preferably an ice tray), and wait. The average American today buys only four bags of ice from the local grocery or convenience store each year. Eighty percent of that amount is sold between Memorial Day and Labor Day.[7] The freezer compartment of the average refrigerator can produce more than enough to meet most people's needs during most of the summer and certainly during the rest of the year. Now you can buy a refrigerator that will actually make your ice for you! Our great-grandparents would be amazed.

Ice cubes became a selling point for early refrigerators precisely because obtaining them (if not making them) had once been so hard. "Time was when you could get a line on a family's social status by the make of their car," explained the *Chicago Tribune* in 1927. "Nowadays you can tell instinctively whether the so and sos are worth knowing by the number of ice cubes per minute their ice box is capable of throwing off."[8] Even though that was probably a joke, it was based on real behavior; however, that kind of competition could only last so long. "Don't freeze more ice cubes than you need," warned *Good Housekeeping* in 1946. "It wastes electricity. Also, you can remove the unneeded ice trays and use this space for storing a few packages of frozen foods for a week or so."[9] Times had changed. Priorities had changed. Being able to make ice cubes in your kitchen was no longer trendy because it had become so common.

As in the early days of the ice industry, melting remains a sign that ice is doing its job—absorbing heat from whatever it was supposed to make cold. Today, some ice cubes have more jobs than they once did. "When I think about creating a piece of ice for a drink or a cocktail, just as a bartender, there's definitely a culinary aspect," explains the James Beard Award–winning "ice chef" Hope Clarke. "I don't want to do it just to cool the drink down or just to look pretty. It had to contribute something as well. Flavor, the dilution rate, you have to think about the way it's interacting with the guest. I personally really enjoy creating an interactive type drink."[10] Managing the exact manner in which and speed at which the ice melts, like they did in the early ice industry, is what makes such drinks "interactive."

These kinds of drinks illustrate how the way that ice gets consumed has changed over time. What today's countless industrial and home freezers have meant is that ice is now prized not for its refrigerating capacity but for less tangible qualities such as its hardness, its slipperiness, or its mouth feel. You can drink a cold soda without ice cubes in it. Just put the bottle or can in the refrigerator before you pour it in your cup. Nonetheless, there remains something special about hearing the clink of ice cubes against the sides of the glass. Maybe it is an echo of the history of an industry that once did so much to give Americans the ability to drink any beverages cold.

ACKNOWLEDGMENTS

In some ways, the acknowledgments for this book could duplicate those for my earlier book, *Refrigeration Nation*, since I had the same large pile of research for this book when I started writing as I did when I finished that one. Nevertheless, I feel obliged to mention a few people again for their additional advice and support. Bernard Nagengast sent me a huge array of primary and secondary materials about the Monitor Top from his private collection, which I made ample use of here. Kenny McKenzie at Colorado State University, Pueblo, remains an interlibrary loan superstar. Perhaps most importantly, JHU Press editor Robert Brugger, before he retired, believed that there was another book worth writing about refrigeration and that I was the guy to write it.

One of the advantages of returning to a subject that you've already published on before is that you can correct some of your oversights as spotted in the reviews. Helen Zoe Veit's very generous review of *Refrigeration Nation* in the *American Historical Review* led to much of what is now chapters three and four here. Staying on the subject of consumption, if Nicola Twilley and Cynthia Graber hadn't done a show on cocktails for their amazing podcast *Gastropod*, I don't think I ever would have thought of covering that subject there. (I had already finished my research on ice cream before they aired their episode on that delicious topic.) Two very helpful anonymous reviewers brought in by Johns Hopkins University Press improved the quality of my analysis immeasurably. So did series editor Richard R. John.

Of all the people who assisted me with the new research I did do for this book, my brother, Daniel Rees, deserves top mention for asking me to write an article for an economics journal with him on the health effects of mechanical refrigeration. Because I agreed, he and our coauthor Mark Anderson made me review all the government reports on ice production, which I made great use of in this work. Wendy Shay deserves special mention for directing me to the Louisan E. Mamer Rural Electrification Administration Papers at the Archives Center in the National Museum of American History. Jonathan Grunert from the Colorado

State University–Pueblo library introduced me to the food and drink collection at Virginia Tech, which enhanced both my food history–related classes and this book in particular. I'd also like to thank whoever made it possible for me to access every database at the University of Wisconsin–Madison's library system through my room at the Lowell Center when I visited there in 2015. Thanks to them, I could finish this book at a time when I was afraid it would never happen.

Lastly, I'd like to recognize all of my students at Colorado State University–Pueblo, going back many years to when I started researching refrigeration. You all put up with my fascination with technology and food history in general and refrigeration in particular, even when you were probably expecting your professor to stick with more conventional subjects. Thanks for listening.

NOTES

PREFACE: Why Ice?

1. Jonathan Rees, *Refrigeration Nation: A History of Ice, Appliances, and Enterprise in America* (Baltimore: Johns Hopkins University Press, 2013), 150.

2. Jonathan Rees, *Refrigerator* (New York: Bloomsbury, 2015), 55.

3. Rees, *Refrigeration Nation*, 89–90.

4. *Salt Lake Herald-Republican*, 19 September 1909, 9.

5. Thomas P. Hughes, "The Seamless Web: Technology, Science, Etcetera, Etcetera," *Social Studies of Science* 16 (May 1986): 281–92. Hughes was writing about the traits of technological networks in general. I am the one applying this general concept to ice.

6. This is the second book that I've written which covers the history of the ice industry in some detail. *Refrigeration Nation* stretches from the beginnings of the industry in 1806, through the transition between ice and mechanical refrigeration as the main source of cold in the United States, all the way down to the present. Because that book covers such a long time span, I was unable to go into much detail with respect to most of the technologies I described. By revisiting the ice-related sections of *Refrigeration Nation*, I can describe all the innovations that made this industry possible in much greater detail and explore the impact of the ice industry on the American diet to a much greater degree than I did in that first work.

7. New England Ice Dealers Association, "The Why Book of Ice," 1931, Jan Bluestein Longone Culinary Archives, Special Collections, University of Michigan, Ann Arbor, MI.

INTRODUCTION: A Museum for the Ice Industry

1. This thought experiment is inspired by Laurel Thatcher Ulrich, *The Age of Homespun: Objects and Stories in the Creation of an American Myth* (New York: Vintage, 2001), 3–8.

2. Henry Hall, "The Ice Industry of the United States . . . ," United States Bureau of the Census, 15 June 1883, 8.

3. Walter R. Sanders, *Ice Delivery* (Chicago: Nickerson & Collins, 1922), 86–89.

4. On the cold chains of today, see Nicola Twilley, "The Coldscape," *Cabinet* 47 (Fall 2012), www.cabinetmagazine.org/issues/47/twilley.php, and her forthcoming book on the same subject.

CHAPTER ONE: How to Harvest Ice

1. Carl Seaburg and Stanley Paterson, *The Ice King: Frederic Tudor and His Circle* (Boston: Massachusetts Historical Society, 2003), 33.

2. *Ice Trade Journal* 3 (March 1880): 5.

3. *Wheeling Daily Intelligencer*, 31 January 1880.

4. *Cold Storage and Ice Trade Journal* 15 (March 1906): 31.

5. Jonathan Rees, *Refrigeration Nation: A History of Ice, Appliances, and Enterprise in America* (Baltimore: Johns Hopkins University Press, 2013), 11.

6. Winfield Thompson, "The Lower Kennebec," *New England Magazine* 17 (February 1895): 677.

7. John Daly Fleming, *Richmond on the Kennebec* (Lewiston, ME: Twin City Printery, 1966), 53.

8. Tom Lewis, *The Hudson: A History* (New Haven, CT: Yale University Press, 2005), 244.

9. *Milwaukee Sentinel*, 15 January 1893, 5.

10. Jennie G. Everson, *Tidewater Ice of the Kennebec River* (Freeport: Maine State Museum, 1970), 30.

11. This entire account is drawn from many available descriptions of the harvesting process, especially Henry Hall, "The Ice Industry of the United States," United States Bureau of the Census, 15 June 1883, 8–9; W. P. Blake, "Mining and Storing Ice," *Journal of the Franklin Institute* 86 (November 1883): 355–69; Everson, *Tidewater Ice*, 44–74; and William T. Wood and Company, "Finest Quality Ice Tools of Every Description," ca. 1885, Winterthur Library, Wilmington, DE.

12. Blake, "Mining and Storing Ice," 356.

13. Ibid., 360.

14. Amos Ives Root, *What to Do, and How to Be Happy While Doing It* (Medina, OH: A. I. Root, 1888), 122.

15. W. T. Wood and Company quoted in Blake, "Mining and Storing Ice," 360.

16. Wood, "Finest Quality Ice Tools."

17. Theron L. Hiles, *The Ice Crop: How to Harvest, Store, Ship and Use Ice* (New York: Orange Judd, 1893), 20.

18. *St. Louis Globe Democrat*, 4 January 1879, 3.

19. Oscar Edward Anderson Jr., *Refrigeration in America: A History of a New Technology and Its Impact* (Princeton, NJ: Princeton University Press, 1953), 107.

20. Blake, "Mining and Storing Ice," 364.

21. *Omaha Bee*, 9 February 1908, 3.

22. Hiles, *Ice Crop*, 35–38.

23. Hall, "Ice Industry," 9.

24. This description of icehouse construction derives from many sources, especially Everson, *Tidewater Ice*, 205–13; Hall, "Ice Industry," 9; and Hiles, *Ice Crop*, 45–53.

25. *New York Tribune*, 29 March 1880, 1.

26. Hall, “Ice Industry,” 21–22.

27. *Washington Evening Star*, 3 August 1901, 18.

28. L. C. Ballard, “Maine Ice Industry,” Fifth Annual Report of the Bureau of Labor Statistics for the State of Maine (Augusta, ME: Burleigh & Flint, 1892), 171.

29. Ibid., 165.

30. *New York Sun*, 17 February 1890, 2.

31. Ballard, “Maine Ice Industry,” 164.

32. Everson, *Tidewater Ice*, 111, 122–23.

33. *Washington Evening Star*, 3 August 1901, 18.

34. Lauretta Wieland, “The Pewaukee Ice Industry: A Tale of a Remarkable Industry,” ca. 1981, 8, 10, Pamphlet Collection, State Historical Society of Wisconsin, Madison, WI.

35. Edward T. O’Donnell, “The Dawn of New York’s Ice Age,” *New York Times*, 31 July 2005, CY3.

36. *New York World*, 27 May 1889, 2.

37. *New York Journal* quoted in Edward P. Kohn, *Hot Time in the Old Town: The Great Heat Wave of 1896 and the Making of Theodore Roosevelt* (New York: Basic Books, 2010), 195.

38. *New York Times* quoted in Gavin Weightman, *The Frozen-Water Trade* (New York: Hyperon, 2003), 237.

39. Kohn, *Hot Time in the Old Town*, 91.

40. Owen Wilson, “The Admiral of the Atlantic Coast,” *World’s Work* 13 (April 1907): 8718–19.

41. Gustavus Myers, *The History of the Great American Fortunes*, vol. 3 (Chicago: Charles H. Kerr, 1910), 294.

42. Richard O. Cummings, *The American Ice Harvests: A Historical Study in Technology, 1800–1918* (Berkeley: University of California Press, 1949), 87–88; Wilson, “Admiral of the Atlantic Coast,” 8718–19.

43. Sterling Beeson, “Prosecuting the Ice Men,” *World To-Day* 11 (September 1906): 941.

44. *Chicago Daily Inter Ocean*, 26 April 1896, 33.

45. *New York Sun*, 9 March 1890, 1.

46. Rees, *Refrigeration Nation*, 64–75.

CHAPTER TWO: How to Manufacture Ice

1. On Gorrie, see Oscar Edward Anderson Jr., *Refrigeration in America: A History of a New Technology and Its Impact* (Princeton, NJ: Princeton University Press, 1953), 74–77; Barry Donaldson and Bernard Nagengast, *Heat and Cold: Mastering the Great Indoors* (Atlanta: American Society of Heating, Refrigerating and Air-Conditioning Engineers, 1994), 119–20, 124; Carroll Gantz, *Refrigeration: A History* (Jefferson, NC: McFarland, 2015), 38–41; and John Gladstone, “John Gorrie, The Visionary,” *ASHRAE Journal*, December 1998, www.ashrae.org/File%20Library/docLib/Public/200362795143_326.pdf.

2. Minna Scherlinder Morse, "Chilly Reception," *Smithsonian*, July 2002, www.smithsonianmag.com/history/chilly-reception-66099329/?no-ist.

3. Gorrie died long before he could make any money from his invention and even longer before ice machines became a mainstay of American life. In 1914, the state of Florida erected a statue of him in the Statuary Hall at the Capitol in Washington, DC.

4. Jonathan Rees, *Refrigeration Nation: A History of Ice, Appliances, and Enterprise in America* (Baltimore: Johns Hopkins University Press, 2013), 36–38.

5. Frick Company, "Eclipse Refrigerating Machinery," 27–28, Refrigerating Catalogs and Miscellaneous, National Museum of American History, Washington, DC.

6. *New York World*, 24 June 1890, 1.

7. Jonathan Rees, "I Did Not Know . . . Any Danger Was Attached: Safety Consciousness in the Early Ice and Refrigeration Industries," *Technology and Culture* 46 (July 2005): 555.

8. *Scientific American* 42 (19 June 1880): 342.

9. Carl Snyder, "A Frost for Ice Dealers," *Omaha Daily Bee*, 4 April 1897, 16.

10. See Rees, *Refrigeration Nation*, 199n5. Leaking refrigerant from the valves was a common problem, not just when ammonia was used as the refrigerant. Carbon dioxide, another common refrigerant during this era, had to be kept at even higher pressure than ammonia.

11. See ibid., 31–32. Ammonia has come back into use for industrial refrigeration in recent decades. Its faults as a refrigerant have been minimized by technological advancements in refrigerating machine construction since the late nineteenth century.

12. Arthur L. Hunt, "Manufactured Ice," *Manufactures, Part III*, Census Reports IX (Washington, DC: United States Census Office, 1902), 680.

13. Rees, *Refrigeration Nation*, 46.

14. W. J. Rushton, "Early Days of the Manufacture of Ice," *Ice and Refrigeration* 51 (November 1916): 152.

15. *Ice and Refrigeration* 7 (September 1894): 174.

16. *Frank Leslie's Illustrated Weekly*, 14 February 1891, 30.

17. *New York Tribune*, 8 March 1903, B2.

18. Donaldson and Nagengast, *Heat and Cold*, 52.

19. *Ice Trade Journal* 20 (December 1896): 6.

20. Hollis Godfrey, "The City's Ice," *Atlantic Monthly* 104 (July 1909): 121.

21. Cyrus Sylvester, "Midsummer Ice Crop," *Fort Worth Morning Register*, 11 July 1897, 5.

22. Hugh B. Meloy, "Manufactured Ice," Department of Commerce and Labor, Bureau of the Census, *Manufactures 1905*, pt. 3 (1908), 45.

23. Elli Morris, *Cooling the South: The Block Ice Era, 1875–1975* (Richmond, VA: Wackophoto, 2007), 17–18.

24. *Transactions of the American Society of Refrigerating Engineers* 2 (3 and 4 December 1906): 216–19.

25. Meloy, "Manufactured Ice," 503.

26. Ibid.; Van Rensselaer H. Green, "New and Improved Arrangements in the Installation and Operation of Ice Factories," *Ice and Cold Storage* 13 (December 1910): 266.

27. Anderson, *Refrigeration in America*, 29.

28. J. Burnitz Bacon, "Lager Beer in America," *Frank Leslie's Popular Monthly* 14 (August 1882): 213–17.

29. *Scientific American* 63 (6 September 1890): 149.

30. Anderson, *Refrigeration in America*, 55.

31. De La Vergne Manufacturing Company, "The De La Vergne Manufacturing Company of New York City," 3rd ed. [catalog] (New York, 1890), 63–66.

32. Donaldson and Nagengast, *Heat and Cold*, 136.

33. *Scientific American Supplement* 28 (2 November 1889): 11536–37.

34. Hugh F. Fox, "The Prosperity of the Brewing Industry," *Annals of the American Academy of Political and Social Science* 34 (November 1909): 56–57.

35. Frick Company, "Eclipse Refrigerating Machines . . . " (1890), 123.

36. *San Francisco Call*, 3 August 1890, 8.

37. Anderson, *Refrigeration in America*, 99. Of course, not all small ice machines made ice. Slaughterhouses, to name one example, wanted refrigerating equipment simply to keep their meat cold.

38. *New York Tribune Illustrated Supplement*, 30 June 1901, 3.

39. Henry Hall, "The Ice Industry of the United States," United States Bureau of the Census, 15 June 1883, 20, 6.

40. Arthur Hunt, "Manufactured Ice," Census Bulletin No. 174, 22 May 1902, 3, 13.

41. Meloy, "Manufactured Ice," 495.

42. S. S. Vandervaart, "Commercial Uses of Refrigerating Machinery," *Ice and Refrigeration* 51 (November 1916): 178–79.

43. *Washington Evening Star*, 12 August 1906, 3.

44. H. D. Pownall, "Raw Water Ice," *Ice and Refrigeration* 49 (December 1915): 302.

45. *New York Sun*, 9 March 1890, 21.

46. Donaldson and Nagengast, *Heat and Cold*, 190.

47. *Electrical World* 62 (6 December 1913): 1161.

48. Louis K. Doelling, "Twenty Five Years' Evolution of Refrigeration," *Ice and Refrigeration* 51 (November 1916): 161.

49. Rees, *Refrigeration Nation*, 53.

50. *Electrical World* 59 (2 March 1912): 482.

51. R. Louis Lloyd, "Electricity in Refrigeration," *Journal of the Franklin Institute* 166 (December 1908): 463.

52. H. D. Pownall quoted in *American Industries* 11 (May 1911): 29.

53. Lloyd, "Electricity in Refrigeration," 461.

54. *Cold Storage* 4 (August 1900): 38.

55. Meloy, "Manufactured Ice," 495.

56. John G. Hawes, "Manufactured Ice," United States Bureau of the Census, *Census of Manufactures*, vol. 2 (1914), 439.

57. Bureau of the Census, "Manufactured Ice," *Fourteenth Census of the United States, Manufactures: 1919* (Washington, DC: US Government Printing Office, 1922), 4, 3.

58. Anderson, *Refrigeration in America*, 107, 109.

59. Hawes, "Manufactured Ice," 3.

60. Ibid.

61. Bureau of the Census, "Manufactured Ice," 956, 955.

62. Natural Ice Association of America, "The Handwriting on the Wall," New York, ca. 1913, Box B-13, Lake Mohonk Mountain House Collection, Hagley Museum and Library, Wilmington, DE.

63. John E. Starr, "The Passing of Natural Ice," *Ice and Cold Storage* 23 (June 1920): 112.

CHAPTER THREE: How Ice (and the Perishable Food It Preserved) Made It to Consumers

1. On the early days of the American ice industry, see Jonathan Rees, *Refrigeration Nation: A History of Ice, Appliances, and Enterprise in America* (Baltimore: Johns Hopkins University Press, 2013), 11–30.

2. Tammis K. Groft, W. Douglas McCombs, and Ruth Greene-McNally, *Hudson River Panorama: A Passage through Time* (Albany: State University of New York Press, 2009), 80.

3. *New York Tribune Illustrated Supplement*, 30 June 1901, 3.

4. *New York Tribune*, 13 December 1891, 16.

5. *New York Tribune Illustrated Supplement*, 30 June 1901, 3.

6. A similar situation existed in Washington, DC, except that it was particularly dependent on ice from Maine. See *Washington Evening Star*, 12 August 1906, 3.

7. Robert Maclay, "The Ice Industry," in *One Hundred Years of American Commerce: A History*, vol. 2, ed. Chauncey M. Depew (New York: D. O. Haynes, 1895), 468.

8. Paul Angle, "Ice Wagons," *Chicago History* 5 (Spring 1959): 250.

9. *Washington Evening Star*, 12 August 1906, 3.

10. Ibid.

11. Walter R. Sanders, *Ice Delivery* (Chicago: Nickerson & Collins, 1922), 197–98.

12. *Washington Evening Star*, 12 August 1906, 3.

13. *Electrical World* 59 (2 March 1912): 482.

14. Carpenter Ice Company, 1882, Ice Cards (Uncatalogued Ephemera), Hagley Museum and Library, Wilmington, DE.

15. *New York Tribune Illustrated Supplement*, 30 June 1901, 3.

16. Paul Angle, "Ice Wagons," *Chicago History* 5 (Spring 1959): 247–48.

17. Ibid., 248–49.

18. Washington Ice Company, 1883, Ice Cards.

19. Sanders, *Ice Delivery*, 48.

20. *New York World*, 27 May 1889, 2.

21. *Washington Evening Star*, 12 August 1906, 3.

22. C. W. Southward, “Is It Desirable to Eliminate the Ice Peddler?,” *Ice and Refrigeration* 32 (April 1907): 193.

23. The exact details of the entire accounting system are described in great detail in Sanders, *Ice Delivery*, 84–121.

24. W. K. Trechsel, “The Coupon System,” *Ice and Refrigeration* 54 (February 1918): 68.

25. *New York Tribune*, 13 December 1891, 16.

26. Laurence Pringle, *Ice! The Amazing History of the Ice Business* (Honesdale, PA: Calkins Creek, 2012), 50.

27. W. J. Rushton, “The Ice Man as Seen by Public,” *Hartford* (KY) *Republican*, 6 August 1909, 2.

28. “How New York Is Fed,” *Scribner's* 14 (October 1877): 735.

29. “Refrigerator and Heater Cars,” *Car Builders' Cyclopedia* (New York: Simmons-Boardman, 1928), 164, 162.

30. Donald L. Miller, *City of the Century: The Epic of Chicago and the Making of America* (New York: Simon & Schuster, 1996), 207–8.

31. Alfred D. Chandler Jr., *The Visible Hand: The Managerial Revolution in American Business* (Cambridge, MA: Belknap Press of Harvard University Press, 1997), 300.

32. Roger Horowitz, *Putting Meat on the American Table: Taste, Technology, Transformation* (Baltimore: Johns Hopkins University Press, 2006), 19, 34.

33. Louis F. Swift, *The Yankee of the Yards: The Biography of Gustavus Franklin Swift* (Chicago: A. W. Shaw, 1927), 191.

34. Susanne Freidberg, *Fresh: A Perishable History* (Cambridge, MA: Harvard University Press, 2009), 71.

35. William Cronon, *Nature's Metropolis: Chicago and the Great West* (New York: W. W. Norton, 1991), 234.

36. Lee E. Lawrence, “The Wisconsin Ice Trade,” *Wisconsin Magazine of History* 48 (Summer 1965): 263.

37. Horowitz, *Putting Meat on the American Table*, 34.

38. *New York Tribune Illustrated Supplement*, 1 April 1900, 14.

39. Charles H. Stevenson, *The Preservation of Fishery Products for Food* (Washington, DC: US Government Printing Office, 1899), 359–61.

40. Ibid., 363.

41. Elli Morris, *Cooling the South: The Block Ice Era, 1875–1975* (Richmond, VA: Wackophoto, 2007), 28.

42. Stevenson, *Preservation of Fishery Products*, 361.

43. Freidberg, *Fresh*, 238.

44. Richard J. Orsi, *Sunset Limited: The Southern Pacific Railroad and the Development of the American West, 1850–1930* (Berkeley: University of California Press, 2005), 329.

45. J. S. Leeds, “Refrigeration of Citrus Fruit in Transit from California,” *Cold Storage and Ice Trade Journal* 36 (October 1908): 54–55.

46. *New York Sun*, 2 September 1894, 9.

47. *San Francisco Call*, 30 August 1892, 3.

48. *Ice and Refrigeration* 66 (May 1924): 449–50.

49. Orsi, *Sunset Limited*, 330.

50. Leeds, "Refrigeration of Citrus Fruit," 54–55.

51. Orsi, *Sunset Limited*, 330–31.

52. Gabriella Petrick, "'Like Ribbons of Green and Gold': Industrializing Lettuce and the Quest for Quality in the Salinas Valley, 1920–1965," *Agricultural History* 80 (Summer 2006): 278.

53. Federal Writer's Project, *Colorado: A Guide to the Highest State* (New York: Hasting House, 1941), 68.

54. Philo K. Blinn, "Development of the Rocky Ford Cantaloupe Industry," The Agricultural Experiment Station at the Colorado Agricultural College, Bulletin 108, March 1906, 13–14; *New York Times*, 22 September 1904, 10.

55. *New York Sun*, 2 September 1894, 9.

56. Robert S. Lynd and Helen Merrell Lynd, *Middletown: A Study in Modern American Culture* (New York: Harvest Books, 1956), 156–57.

57. Ibid. Since mechanical refrigeration did not reach railway cars until the 1950s and there were only eleven electric refrigerators in the whole town when the Lynds visited, this change in diet depended a great deal on the increased use of ice for transport and home storage. However, cold storage using mechanical refrigeration first took off after World War I, and the Lynds attribute some of this change to that too. See Rees, *Refrigeration Nation*, 97; and Lynd and Lynd, *Middletown*, 172.

CHAPTER FOUR: How Ice Changed the American Diet and American Life

1. Kendra Smith-Howard, *Pure and Modern Milk: An Environmental History since 1900* (New York: Oxford University Press, 2014), 5, 15, 18.

2. *El Paso Herald*, 11–12 December 1915, 7.

3. Smith-Howard, *Pure and Modern Milk*, 12.

4. Alan Czaplicki, "'Pure Milk Is Better Than Purified Milk': Pasteurization and Milk Purity in Chicago, 1908–1916," *Social Science History* 31 (Fall 2007): 411.

5. Before the development of the electric household refrigerator, what we now call iceboxes were generally called refrigerators. The term *icebox* became popular during the 1920s in order to distinguish the older technology from this newer one. For much more on iceboxes, see Jonathan Rees, *Refrigeration Nation: A History of Ice, Appliances, and Enterprise in America* (Baltimore: Johns Hopkins University Press, 2013), 119–39.

6. Helen Veit, "An Economic History of Leftovers," *Atlantic*, 7 October 2015, www.theatlantic.com/business/archive/2015/10/an-economic-history-of-leftovers/409255/.

7. Anna Barrows, "The Housekeeper's Vacation," *Good Housekeeping* 57 (July 1913): 109.

8. *Refrigerating World* 48 (July 1914): 49.

9. W. T. Sedgwick, "Ice and the Public Health," *Ice and Refrigeration* 41 (October 1911): 134.

10. Charles S. Mahoney, *The Hoffman House Bartender's Guide* (New York: R. K. Fox, 1905), 117.

11. *Puck*, 15 February 1882, 376.

12. David Wondrich, *Imbibe!*, updated and rev. ed. (New York: Perigee, 2015), 51, 96.

13. *Milwaukee Sentinel*, 27 July 1884, 7.

14. Dale DeGroff, *The Craft of the Cocktail* (New York: Clarkson Potter, 2002), 17.

15. Wondrich, *Imbibe!*, 47–48, 65.

16. Harry Johnson, *The New Illustrated and Improved Bartender's Manual, or How to Mix Drinks of the Present Style* (New York, 1888), 31.

17. *Milwaukee Daily Sentinel*, 26 June 1876, 3.

18. *St. Louis Globe Democrat*, 10 December 1882.

19. *Ice and Cold Storage* 8 (February 1905): 34.

20. Wondrich, *Imbibe!*, 47, 51–52.

21. "Here's How: A Handbook of Recipes of Spiritous and Non-spiritous Drinks Gathered from Authoritative Sources," 1900, 17–19, History of Food and Drink Collections Rare Books, Virginia Tech Special Collections Online.

22. Andrew F. Smith, *Drinking History: Fifteen Turning Points in the Making of American Beverages* (New York: Columbia University Press, 2013), 132, 128.

23. Marilyn Powell, *Ice Cream: The Delicious History* (New York: Overlook, 2005), 8–9.

24. Jeri Quinzio, *Of Sugar and Snow: A History of Ice Cream Making* (Berkeley: University of California Press, 2009), 64, 95–98.

25. *St. Paul* (Minnesota) *Daily Globe*, 11 August 1889.

26. R. Louis Lloyd, "Electricity in Refrigeration," *Journal of the Franklin Institute* 166 (December 1908): 465.

27. *Scientific American* 71 (21 July 1894): 40.

28. Joseph H. Hart, "The Application of Mechanical Refrigeration to Ice Cream Manufacture," *National Provisioner* 34 (November 1906): 397, 402–3.

29. *New York Sun*, 2 August 1891, 2.

30. Anne Cooper Funderburg, *Chocolate, Strawberry and Vanilla: A History of American Ice Cream* (Bowling Green, OH: Bowling Green State University Popular Press, 1995), 64–65.

31. *St. Paul* (Minnesota) *Daily Globe*, 11 August 1889.

32. Laura B. Weiss, *Ice Cream: A Global History* (London: Reaktion Books, 2011), 42.

33. Smith-Howard, *Pure and Modern Milk*, 100–101.

34. See Rees, *Refrigeration Nation*, 111. I devoted a chapter to cold storage in *Refrigeration Nation*, but I only discovered the materials on cold storage facilities that used ice cited below after that book was published. In other words, everything that follows in this chapter is new.

35. Madison Cooper, "Ice Cold Storage," 1901, 14–15, 18, Madison Cooper Papers, Box 2, Smithsonian Archive Center, National Museum of American History, Washington, DC.

36. *Scientific American* 71 (22 September 1894): 182.

37. Cooper, "Ice Cold Storage," 26.

38. Madison Cooper to the Editor of the *Egg Reporter*, *Egg Reporter*, July 1902, Clipping in the Cooper Papers, Box 2.

39. Cooper, "Ice Cold Storage," 5.

40. Madison Cooper, "The Cooper Systems of Refrigeration," ca. 1902, Cooper Papers, Box 2.

41. Madison Cooper, *Practical Cold Storage* (Chicago: Nickerson & Collins, 1904), 15.

42. Alice F. Peterson, "Food Preserved by Cold Storage," *Current Literature* 18 (December 1895): 507.

43. *Colman's Rural World*, 3 December 1902, 1.

44. Cooper, *Practical Cold Storage*, 19.

45. *Milwaukee Sentinel*, 28 March 1897, 21.

46. Rees, *Refrigeration Nation*, 109.

47. Nigel Brown, *Ice-Skating: A History* (New York: A. S. Barnes, 1959), 81.

48. Marc Ferris, "Ice Skating," *Encyclopedia of the City of New York*, ed. Kenneth T. Jackson (New Haven, CT: Yale University Press, 1995), 579.

49. "Skating," *Appleton's Dictionary of New York and Its Vicinity*, 6th ed. (New York: D. Appleton, 1880), 191.

50. Carroll Gantz, *Refrigeration: A History* (Jefferson, NC: McFarland, 2015), 61; Brown, *Ice-Skating*, 140–41.

51. *Scientific American*, 21 February 1880, 113.

52. *New York Times*, 6 November 1896.

53. *New York Sun*, 5 January 1896, 2.

54. *Milwaukee Sentinel*, 28 March 1897, 21. Since ice skates cut into ice, they leave grooves on the surface of the rink. The ice once in that groove becomes snow on the surface, which impedes other skaters. This explains why resurfacing after large crowds would have been necessary if the refrigerating machinery hadn't been as good.

55. *New York Sun*, 5 January 1896, 2.

56. *Raleigh News-Observer-Chronicle*, 14 July 1893.

57. Rees, *Refrigeration Nation*, 31–32.

58. Paul Charles Simonini and Robert Judson Kirkpatrick, "Design and Investigation of Cost of an Artificial Ice Skating Rink" (BS thesis, College of Engineering, University of Illinois, 1911), 3–4, 14.

59. Martin R. Carpenter, "Building Ice-Skating Rinks," *Refrigerating Engineering* 21 (May 1931): 334.

60. C. F. Greeves-Carpenter, "Ice Rinks Invade the World," *Refrigerating Engineering* 37 (February 1939): 1.

61. Carpenter, "Building Ice-Skating Rinks," 337, 339.

62. Bernard Nagengast, "A History of Comfort Cooling Using Ice," *ASHRAE Journal*, February 1999, 50; available at www.ashrae.org/about-ashrae/ashrae-and-industry-history.

63. Gail Cooper, *Air-Conditioning America: Engineers and the Controlled Environment, 1900–1960* (Baltimore: Johns Hopkins University Press, 1998), 9.

64. Nagengast, "History of Comfort Cooling," 50.

65. *New York Tribune*, 17 May 1896, 28.

66. *New York Times*, 9 January 1910, X9.

67. *New York Tribune*, 28 June 1914, 8.

68. "How Ice Is Manufactured," *Current Literature* 5 (August 1890): 114.

69. Nagengast, "History of Comfort Cooling," 53.

70. Cooper, *Air-Conditioning America*, 90–91, 110.

CHAPTER FIVE: How Household Refrigerators Changed the Ice Market Forever

1. Robert Maclay, "The Ice Industry," in *One Hundred Years of American Commerce: A History*, vol. 2, ed. Chauncey M. Depew (New York: D. O. Haynes, 1895), 469.

2. J. F. Nickerson, "The Household Machine as Competitor of Ice," *Refrigeration* 37 (July 1925): 34.

3. Ruth Schwartz Cowan, *More Work for Mother* (New York: Basic Books, 1983), 135.

4. Carroll Gantz, *Refrigeration: A History* (Jefferson, NC: McFarland, 2015), 111.

5. Barry Donaldson and Bernard Nagengast, *Heat and Cold: Mastering the Great Indoors* (Atlanta: ASHRAE, 1994), 225.

6. Jonathan Rees, *Refrigeration Nation: A History of Ice, Appliances, and Enterprise in America* (Baltimore: Johns Hopkins University Press, 2013), 150–52.

7. General Electric, "Now Comes—Simplified Electric Refrigeration," advertisement, 1927, Bernard Nagengast Collection.

8. Donaldson and Nagengast, *Heat and Cold*, 225.

9. Carroll M. Gantz, *Design Chronicles: Significant Mass-Produced Designs of the 20th Century* (Atglen, PA: Schiffer, 2005), 69.

10. General Electric, "An Amazing Record," *Refrigerating Engineering* 18 (December 1929): 51.

11. GE quoted in Gantz, *Refrigeration*, 112.

12. Cowan, *More Work for Mother*, 136, 138.

13. David Nye, *Electrifying America: Social Meanings of a New Technology, 1880–1940* (Cambridge, MA: MIT Press, 1990), 276.

14. Gantz, *Refrigeration*, 143.

15. General Electric, "Retail Salesman's Handbook on General Electric Refrigerators," ca. 1928, 44, 43.

16. Ibid., 41, 43.

17. General Electric, "Food: Its Proper Refrigeration," April 1926, 5. Many thanks

to Bernard Nagengast for sending me a scan of this and other pamphlets from his personal collection for citation in this chapter.

18. Ibid.

19. General Electric, "Simplified Electric Refrigeration," Cleveland, OH, 1927, 6, 20, Bernard Nagengast Collection.

20. General Electric, "Save Time, Energy and Money—Don't Market Every Day," ca. 1927, Refrigerator Advertisement Collection, ASHRAE Library, Atlanta, GA.

21. General Electric, "Food," 8.

22. General Electric, "Retail Salesman's Handbook," 17.

23. General Electric, "3 Year Guarantee," ca. 1930, Refrigerator Advertisement Collection.

24. General Electric, "Retail Salesman's Handbook," 40.

25. Warren M. Persons, "An Economist's Appraisal of Domestic Refrigeration," The Temperature Research Foundation, 2, 5, National Museum of American History, Refrigeration and Miscellaneous Pamphlets, Washington, DC. These sales figures are slightly different from the ones I cite in *Refrigeration Nation* (164) because they come from a different, earlier source. Nonetheless, both sets of figures show a huge spike over this same period of time.

26. Halbert H. Hall, "The Ice Industry Today," *Ice and Refrigeration* 72 (May 1927): 413.

27. John Nickerson, "Ice Finance," *Ice and Refrigeration* 71 (December 1926): 437–38.

28. Rees, *Refrigeration Nation*, 163.

29. Horace B. Drury, "Production and Capacity Control in the Ice Industry under the NRA," National Recovery Administration, Trade Practices Study Section, 1936, 10.

30. *Ice and Refrigeration* 70 (May 1926): 534.

31. Ibid.

32. *Ice and Refrigeration* 71 (August 1926): 89.

33. Edmund Copeland quoted in Donaldson and Nagengast, *Heat and Cold*, 226, 228.

34. Oscar Edward Anderson Jr., *Refrigeration in America: A History of a New Technology and Its Impact* (Princeton, NJ: Princeton University Press, 1953), 216.

35. Elli Morris, *Cooling the South: The Block Ice Era, 1875–1975* (Richmond, VA: Wackophoto, 2007), 58, 75.

36. Anderson, *Refrigeration in America*, 217.

37. Bernard Nagengast, "A History of Comfort Cooling Using Ice," *ASHRAE Journal*, February 1999, 54–55; available at www.ashrae.org/about-ashrae/ashrae-and-industry-history.

38. General Electric, "An Amazing Record," 51.

39. William L. Holladay, the first Monitor Top salesman, later listed 27 different models sold over the Monitor Top's ten-year production run, with different sizes and even different refrigerants. See William L. Holladay, "The Monitor Top Refrigerator," *ASHRAE Journal*, September 1994, 54–55. Courtesy of Bernard Nagengast.

40. General Electric, "Dependable Performance—the True Measure of Value," *Time* 18 (26 October 1931): 5.

41. Gantz, *Refrigeration*, 148.

42. Andrew D. Althouse and Carl Turnquist, *Modern Electric and Gas Refrigeration*, 2nd ed. (Chicago: Goodheart-Wilcox, 1936), 104–8.

43. Gantz, *Refrigeration*, 159–65.

44. Ronald Tobey, *Technology as Freedom: The New Deal and the Electrical Modernization of the American Home* (Berkeley: University of California Press, 1996), 123.

45. William J. Enright, "Refrigerator Men to Push Specials," *New York Times*, 26 November 1939, 64.

46. This gave GE consumers the opportunity for the first time to store things on top of their refrigerators. While this may not seem like a particularly strong selling point, even little improvements mattered in a very competitive market.

47. Not only does the Monitor Top remain the most recognizable antique refrigerator today, because of its machinery on the top, but (at least according to the *New York Post*) it is also "America's oldest continually operating fridge." "Its basketball-like top houses the mechanical assembly which is still cooling food as though it was bought yesterday," the newspaper reported in 2013. People interested in restoring Monitor Tops can still find parts and instructions on the Internet. See *New York Post*, 5 July 2013, http://nypost.com/2013/07/05/americas-oldest-fridge-still-keeping-cool/.

48. E. B. Newill, "Improvements in Household Refrigerators," *Refrigerating Engineering* 34 (July 1937): 24.

49. Arthur J. Donniez, "Old Refrigerators: What They Haven't Got," *Good Housekeeping*, June 1940, 145, 184.

50. Donaldson and Nagengast, *Heat and Cold*, 228–29.

51. Holladay, "Monitor Top Refrigerator," 54.

52. Gantz, *Refrigeration*, 150–52.

53. Donniez, "Old Refrigerators," 145.

54. Newill, "Improvements in Household Refrigerators," 24.

55. Ernest Freeberg, *The Age of Edison: Electric Light and the Invention of Modern America* (New York: Penguin, 2013), 298.

56. Nye, *Electrifying America*, 276.

57. David L. Fiske, "The Refrigeration Industry in the United States," *Refrigerating Engineering* 43–44 (9 June 1942), offprint in possession of author.

58. Tobey, *Technology as Freedom*, 156.

59. Cowan, *More Work for Mother*, 139.

60. *Chicago Tribune*, 15 February 1942, 5.

61. Anderson, *Refrigeration in America*, 221.

62. Rees, *Refrigeration Nation*, 164.

63. US Department of Agriculture, Bureau of Home Economics, "The Household Refrigerator: Points to Look for in Selecting," ca. 1946, 2, Louisian E. Mamer

Rural Electrification Administration Papers, 1935–94, Box 11, Collection No. 862, National Museum of American History, Archives Center, Washington, DC.

64. US Department of Agriculture / Rural Electrification Administration, Cooperatives Operations Division, "Refrigeration Instruction Booklet," May 1941, Mamer Papers, Box 11.

65. H. L. Andrews, "Future Demands for Refrigeration and Air Conditioning," *Refrigerating Engineering* 47 (January 1944): 14.

66. US Department of Agriculture, "How to Choose and Use Your Refrigerator," October 1946, 8, Mamer Papers, Box 11.

67. Mrs. Raymond Ricketts, "Never Underestimate the Power of Electricity," *Rural Electrification News* 17 (April–May 1952): 14.

68. J. Kenneth Whitteker, "The Ice Man Goeth," *Boating* 7 (June 1960): 140.

69. Morris, *Cooling the South*, 82, 89.

70. Ibid., 91; Ian Parker, "The Emperor of Ice," *New Yorker*, 12 February 2001, 62, 69.

CONCLUSION: The Inevitability of Melting

1. W. P. Blake, "Mining and Storing Ice," *Journal of the Franklin Institute* 86 (November 1883): 367.

2. Walter Sanders, *Ice Delivery* (Chicago: Nickerson & Collins, 1922), 293.

3. Thomas P. Hughes, *Networks of Power: Electrification in Western Society, 1880–1930* (Baltimore: Johns Hopkins University Press, 1983), 6.

4. Ibid., 21.

5. Thomas P. Hughes, *American Genesis: A History of the American Genius for Invention* (New York: Penguin, 1989), 72.

6. New State Ice Co. vs. Liebmann, 285 U.S. 262 (1932).

7. Guy Trebay, "I Like My Ice Chilled Just So," *New York Times*, 8 August 2008, www.nytimes.com/2008/08/10/fashion/10ice.html?pagewanted=all&_r=0.

8. W. E. Hill, "Artificial Ice," *Chicago Tribune*, 27 November 1927, 7.

9. Mary Dobbs, "What Your Refrigerator Can Do," *Good Housekeeping* 122 (January 1946): 105.

10. Eric Sandler, "Houston's Own Ice Queen: Frozen Bar Wiz Aims to Change the Way You Drink Forever," *CultureMap Houston*, 4 April 2015, http://houston.culturemap.com/news/restaurants-bars/04-02-15-houstons-own-ice-queen-frozen-bar-wiz-aims-to-change-the-way-you-drink/.

INDEX